Advanced Router Architectures

Advanced Router Architectures

Axel K. Kloth

Taylor & Francis
Taylor & Francis Group

Boca Raton   London   New York

A CRC title, part of the Taylor & Francis imprint, a member of the
Taylor & Francis Group, the academic division of T&F Informa plc.

Published in 2006 by
CRC Press
Taylor & Francis Group
6000 Broken Sound Parkway NW, Suite 300
Boca Raton, FL 33487-2742

No claim to original U.S. Government works
Printed in the United States of America on acid-free paper
10 9 8 7 6 5 4 3 2 1

International Standard Book Number-10: 0-8493-3550-7 (Hardcover)
International Standard Book Number-13: 978-0-8493-3550-1 (Hardcover)
Library of Congress Card Number 2005052895

Library of Congress Cataloging-in-Publication Data

Kloth, Axel K.
   Advanced router architectures / Axel K. Kloth.
      p. cm.
   Includes bibliographical references and index.
   ISBN 0-8493-3550-7 (alk. paper)
   1. Routers (Computer networks) 2. Computer network architectures. I. Title.

TK5105.543.K57 2006
004.6--dc22   2005052895

Taylor & Francis Group
is the Academic Division of Informa plc.

Visit the Taylor & Francis Web site at
http://www.taylorandfrancis.com

and the CRC Press Web site at
http://www.crcpress.com

Preface

The Internet has become a part of our daily lives. Routers, switches and transmission equipment make up the backbone of the Internet—yet many users and even data communications and telecommunications service technicians do not understand how these nodes really work.

This book addresses everyone who is generally interested in advanced router architectures, and in particular, how the components of advanced routers work together and how they are integrated with each other. It also explains the background behind why these building blocks perform certain functions, and how the function is implemented in general.

It does not replace the wide variety of very good books that exist today and address each of the items in much more depth, but lack the "glue" to fit them all together. While this book tries to tie it all together, it does not go into particular depths in any of the chapters. This is intended to be an introduction that triggers a deeper interest towards the subject matter and hopefully leads to a more in-depth interest by the reader. While this book aims at a general understanding of the issues, it is not intended to provide a register- or bit-level discussion of functions. An in-depth analysis of the enqueuing and dequeuing algorithms in a traffic manager can be found in books that specialize on schedulers for traffic management. This book explains why traffic management may be important in certain applications, what the traffic manager does, and how it connects to the rest of the router. It also explains what some implications are of the introduction or omission of a traffic manager into an advanced router, but it will not provide guidance as to what the memory sizes, the latency or the word lengths should be. The same is true for any other subject, be it switch fabrics, network processors on the line cards, or framers and MACs on the port cards; this book presents an overview of the functions, why they are performed, and how in general they are implemented. It does not discuss the issue of the optimum scheduler, the ultimate fairness in scheduling, or the frame format of T3/E3/J3. This can be found elsewhere, either in the ITU-T standards or in the many good books describing these formats.

This book, however, will in general explain which mandatory and which optional building blocks can be found in an advanced router, and how these building blocks work in conjunction to ensure that the Internet works in the way we have come to expect it.

The Author

Axel Kloth is the founder and CTO of Parimics, Inc., a fabless semiconductor manufacturer focused on chipset architectures for image analysis. Axel has more than 12 years of experience in high performance computing, networking, and low latency switching as well as experience in architecting and implementing large chips. He has authored or coauthored multiple patents. Prior to Parimics, he worked for Siemens' Public Networks Group and Siemens Microelectronics — later Infineon Technologies — and HotRail and subsequently Conexant and Mindspeed Technologies.

Additionally, he has written several books and technical articles, covering basic technologies and applications and deployments of these. Axel was on the technical advisory boards of HotRail, Z-Force, and Zedix. Axel finished his post-graduate studies in physics and information technologies at CAU Kiel in Germany. His diploma thesis covered what has become the basic technology in Xenon (HID) headlights, and short pulse as well as high energy laser physics. High performance computing, especially MPP SIMD architectures, combined with the low latency switching and interconnect knowledge, contributed to the unique technology behind Parimics' chipset.

Table of Contents

1 Introduction

The basic Internet Protocol version 4 (IPv4) router described in RFC1812 has performed well so far, and survived much longer than its estimated lifetime. The initial drafts of the Internet assumed a maximum of around 20,000 nodes, and it was never intended to exceed that number. The basic premise for the Internet was to survive a massive nuclear attack, and therefore the topology of the Internet was planned to be a mesh of nodes. All of these nodes were intended to be separate and independent, serving only a few users each. Cost of operation and cost per bit transmitted were not an issue, because the Department of Defense (DoD) ran it.

Then the Internet was opened to government agencies, to schools and universities, and finally to the public. Internet Service Providers (ISPs) offered access to the Internet, and the various users generated increasing amounts of traffic. This was becoming too expensive to be operated by the DoD, and therefore it was privatized. All of a sudden, the rules changed. The Internet was subject to the basic logic of demand and supply, to economic requirements. It needed to provide revenue and generate enough profit by itself to run and upgrade the network.

The Internet therefore has changed dramatically, and so have its requirements. Not only are there more nodes, servers, and clients, but more importantly, the Internet has become an integral part of data communications. It now is used to run even mission-critical applications across it. Customers have grown used to it, and now all of a sudden they request Service Level Agreements (SLAs). These SLAs define a minimum System Uptime, Accessibility, Availability, and guaranteed throughput in terms of bits per second and maximum latency for certain traffic classes that may or may not be associated with certain customers or customer groups. While in the beginning only the US and Europe participated in the Internet, now China and India have joined the ranks. This larger number of simultaneous users calls for sustained address uniqueness, therefore requiring a larger address space. This has led the architects of the Internet to rethink a few items, and the outcome was Internet Protocol next generation (IPng), later renamed to Internet Protocol version 6 (IPv6).

To make it perfectly clear, based on the assumptions the architects made for the Internet, there was nothing wrong with the Internet. Based on that, the first routers were built using current and obsolete computers; there was nothing wrong with that approach, either. All devices fulfilled what they were intended to do. Since the very first routers used the telephony network (Plain Old Telephony Services, POTS) as a backbone and for the access to the data network, their data rates and throughput were limited by the capabilities of the modems to which they were connected. The first modems did not deliver more than 1200 bit/s unidirectional. Therefore, the routers had to deal with a maximum throughput of 150 byte/s; they were clearly neither input/output (I/O)- nor compute-bound. That changed over time since increasing processing of packets was required. Encapsulation,

1

translation, routing, and access control became part of the feature set, and later information was added about routing tables and their mutual updates. All these features required more compute power, but since they were internal, they seemed not to impact the router itself. That was and is not entirely true. In spite of limited processing power in the first Central Processing Units (CPUs), the very first routers were I/O bound; whereas the later generations of routers became compute-bound because of the growth of the number of protocols that had to be supported, and because of a very limited increase in line rates—due to the fact that the Time Division Multiplex (TDM)-based backbone of communications was only capable of supporting DS0 and T1/E1 line rates. Routers always have had a line-specific logic part, an adaptation layer between the line-specific logic and the core, higher layer processing and routing code, some switching logic in the core, and a generic control unit for administration. These functions can be implemented and built separately and then integrated into a chassis, or they can be integrated into a single board. They can be implemented in hardware, in software, or in a mixture of both. More so, the new generation of IPv4 routers was built to the then-current expectations and requirements. The newer routers were built around a bus or a simple crosspoint switch with a centralized arbiter to interconnect the line cards, and had a very basic Operation, Administration, Maintenance and Provisioning (OAM&P) software running somewhere in the centralized Operating System (OS). These routers were either single board ("Pizza Box") or modular designs housed in a chassis with either a backplane or a midplane.

However, since the Internet and its use has changed so dramatically, the old IPv4 routers just do not fulfill the requirements any more—even if they were architected today. This book will explain what changed, what the impact on the router's internal architecture is, and what needs to be changed to overcome that problem. Since the prerequisites have changed, the new generation of routers for IPv6 must be different from the old routers.

A huge change has started to take place already. While in the beginning the Internet used the Public Switched Telephony Network (PSTN) Tele Communications (TelCo) infrastructure, now the TelCo industry has started to use the Internet infrastructure as a backbone. This change is fundamental. In the early days, Class 5 Central Office (CO) switches were the concentrators for end users with dial-up ports and modems, and they connected to a hierarchy of Class 4 CO switches through T1/E1 or higher multiplexed digital hierarchies (Plesiochronous Digital Hierarchy, PDH). In fact, the Class 4 switches were the backbone of the Internet. Now the Internet is the backbone for the TelCo industry, the carriers, the Regional Bell Operating Companies (RBOCs) and Competitive Local Exchange Carriers (CLECs), and the datacom industry. It is the core of the integrated, unified, and converged network. The Class 5 CO switches are not the concentrators any more—they have to provide Digital Susbscriber Line (DSL) ports for high-bandwidth users, and the DSL ports are not routed to the Class 5 switch core anymore. They are physically and logically separated and connected to Asynchronous Transfer Mode (ATM) switches or Digital Subscriber Line Access Multiplexers (DSLAMs). Both the PDH uplinks from the Class 5 switch and the DSLAM ports now connect to carrier-grade routers instead of Class 4 CO switches.

Current IPv4 routers deal with IPv4 packets. An IPv4 packet is a datagram of variable length between 40 byte long for Packet over SONET (PoS) and 1518 byte long; for the so-called jumbo frames it may even be 9018 bytes long. Additionally, the line rates are not very high in typical IPv4 routers. Most IPv4 routers top out at 1 Gigabit per second (Gbit/s) per port. Achieving a port throughput of 1 Gbit/s is not too complicated today, neither on the line card, nor within the switch. If additionally averaging the size of the packet, and not considering the worst-case scenario can limit the packet rate, then it is fairly simple. Assuming the average packet size exceeds 1000 bit (125 byte), then the router's ports will have to deal with one million packets per second. If one lookup per packet must be performed because the destination IP address can directly be mapped to an egress port, then the port card must perform one million lookups per second. Content Addressable Memory (CAMs) that achieve these lookup speeds are readily available and cheap. However, current routers are not optimized to deal with different traffic types that may have different real-time requirements or limits of the latency or delay variation. That leaves us with the necessity of modern, advanced routers, even if ATM switching may be the underlying technology in the network or the router itself. With GbE and 10GbE over "dark fiber" catching up, some of the ATM and Synchronous Digital Hierarchy and Synchronous Optical Network (SDH/SONET) infrastructure in the Access and the Metropolitan Area Network (MAN) rings may become obsolete, but in the Wide Area Network (WAN), SDH/SONET will continue to dominate the transport infrastructure, on top of Dense Wavelength Division Multiplex (DWDM) or other technologies to increase the throughput though and over individual fibers. I assume DWDM is not the end of the effort to squeeze in more transport capacity. I assume Quad Phase Shift Keying (QPSK) or Quaternary or Quadrature Amplitude Modulation (QAM) or additional new technologies will come up there. More importantly, these transport infrastructures will change in a way that allows the provider of these services to set up connections fast—basically on demand. Now that raw bandwidth is rather cheap, the most important issue the backbone and service providers face is to provision these fat pipes. Assume there will be a signal—either directly originating from the customer or from the customer to the Network Management Server (NMS) in the Network Management Center (NMC) and then into the providing nodes—to set up connections as needed. Add-Drop-Multiplexers (ADMs) will have to be able to understand that kind of signaling. That will impact routers and ATM switches as well. While the RBOCs and CLECs will be able to make use of this new, cheaper infrastructure in the backbone, cell phone providers especially will benefit from a conversion of the backbone based on Class 4 CO switches into a packet-based network backbone. The cell phone providers' signaling traffic (from roaming and other metadata provided for the cell phone infrastructure) desperately needs an infrastructure that is more cost-effective than the current one (see Figure 1.1)

The cell phone network currently is a portion of the PSTN, but the transition that the cell phone network undergoes—with EDGE and General Packet Radio Service (GPRS), and with 3G—is so tremendous that it will be using the Internet Core Network as a backbone even before the long-distance providers will adopt that same transition.

As a result, routers deployed in the core network today will have to be able to transport PSTN and cell phone traffic tomorrow. Extending existing networks with the same equipment used years ago, therefore, is not good enough. The new situation will look much closer to Figure 1.2.

In short, we can say that the Internet was fine the way it was first conceived and implemented. The requirements changed with more users and different traffic types. Privatization of the Internet core led to another set of new requirements. With the carriers taking over again and a consolidation of carriers and ISPs, a convergence of the separate networks towards one unified network was inevitable. The necessity to make a profit off the Internet and all related and connected networks added the necessity to be able to bill for services depending on their Service Level Agreement (SLA). This had a profound impact on routers, ATM switches, and all other networking gear. The most fundamental impact was on the router design criteria, since best-effort approaches were replaced by SLA guarantees, and the lack of policing and billing was replaced by per-connection policing and billing. Internetworking with other networks was ignored before, and in the future advanced routers will have to be able to interconnect and communicate with all networks in existence

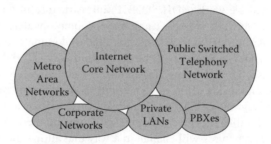

FIGURE 1.1 Current network interconnects.

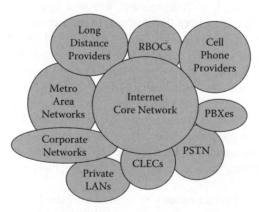

FIGURE 1.2 Future network interconnects.

today. The more protocols are supported by advanced routers or their gateways, the broader the deployment range of these routers and gateways.

CONCLUSION

As a summary, we can safely say that the Internet—as originally described, conceived and implemented—performed its function perfectly. The growing number of users and the requirement to carry different traffic types led to the privatization of the Internet, which in turn meant that it had to be made profitable. With the carriers initially only taking over parts of the backbone of the Internet, and with consolidation in the industry, convergence became a requirement of the operators of the Internet, not only for the users. The impact on routers, ATM switches, and all other networking gear was tremendous. As a result, the design criteria for routers have changed significantly. As a side note, it is important to stress that the need for interconnectivity has reached servers; blade servers now incorporate router blades.

We can conclude that a modern router will have to seamlessly and reliably do the following:

- Losslessly switch and route datagrams
- Be able to perform Segmentation And Reassembly (SAR)
- Perform policing at line speed for thousands of virtual connections simultaneously and thereby enforce and support SLAs
- Perform Traffic Management by queuing and buffering traffic according to SLAs, and drop excess traffic
- Support Traffic Engineering by means of collecting Statistic Traffic Data
- Support a mixture of hierarchical and mesh interconnect infrastructure in terms of data traffic
- Support an overlay network for metadata and signaling data
- Enable traffic rerouting at the edge and provide redundant fail-safe systems towards the core
- Communicate securely within the components of the router
- Communicate securely between the OAM&P card and a Billing Center
- Communicate securely between the OAM&P card and a PKI Center for authentication
- Communicate securely between the OAM&P card and a Network Management Center
- Communicate with the PSTN infrastructure

This will enable the Internet to continue to grow, and the nodes to scale near-linearly.

The components and their interactions determine the routers' deployment range, and price range. We will see in this book which components of a router impact the desired functions.

Specifically, the components and the interaction between the components of a router can be understood as a complex system that, by itself, can perform a wide variety of datapath functions, but must be seen as part of a larger system—the network itself.

The chapters are arranged such that they follow the train of thought as to why the requirements for routers changed, and what had to be done to make them compliant with current necessities. They are ordered as follows:

- Internet Topology Change
- The Carrier Business Model
- Advanced Routers in Central Office Applications
- Function Split
- High Availability (HA)
- The Chassis
- Line Card Functions
- Switch Fabrics
- Operation, Administration, Maintenance and Provisioning (OAM&P)
- Glossary
- Literature

2 Internet Topology Change

OVERVIEW

When the Internet was conceived—then as the ARPAnet—it was intentionally set up to support a mesh interconnect infrastructure with a very flat hierarchy. This was very advantageous for what it was intended to do: survive the outage of a significant number of nodes in case of a nuclear attack. As a result, everything was set up to be "self-routing," and a centralized NMC was not only undesirable, it was unwanted. A centralized Network Management Center (NMC) for signaling and call and transmittal routing would have made ARPAnet vulnerable even if the router network were distributed. Therefore, even if the transport network had been able to transmit datagrams, an outage of the centralized NMC would have rendered useless the transport network's distributed nature and its virtual invulnerability.

As a result, ARPAnet—and consequentially the Internet—has no centralized NMC. However, one has to take into account that ARPAnet was designed to support a maximum of about 20,000 nodes. The architects of ARPAnet never took into consideration such a growth in numbers of nodes, and it is a testament to the genius of the Internet's inventors that it still works. While some of the original design flaws have been fixed, the Internet as of today still does not have a structure that can guarantee fulfillment of future requirements of Internet users. Today, the Internet is a mixture of a mesh and a partially hierarchical structure. In effect, the path of any given packet cannot be predicted. As a result, cell or packet delay variation cannot be predicted, either.

CURRENT STATUS

One of the many implications of the Internet's growth was its conversion from a mesh to a mixture between a mesh and a hierarchical architecture with super-nodes. Although not many people want to hear this, everyone involved in network topologies has to acknowledge that the Internet evolved into an architecture that is very similar to the "old" telephony architecture of the PSTN. For cost reasons, the full mesh was abandoned. The mesh is too expensive for large numbers of users and nodes, and it requires routers with very large numbers of interfaces. A router with large numbers of interfaces and rather low throughput on each of these interfaces is expensive to maintain and requires large routing tables. On top of that, it will require a lot of signaling information to continuously update its routing tables. Therefore, a significant portion of the available bandwidth will be consumed for signaling information and network internal data. This problem is only going to get worse, since the number of interconnects N—logical or physical—goes with $N*(N-1)/2$, and therefore is an N^2 problem. No network operator can afford an N^2 communications protocol on a large number of users or nodes, and therefore the Internet topology had to evolve into a hierarchical structure (see Figure 2.1) with super-nodes. This is not bad news, but it has profound implications of routers or nodes.

While the full mesh very apparently has no overhead in the network and no layers of switches to forward data from the source to the destination, we can see that each node has a large number of interconnects that must be served (see Figure 2.2). This must be done within the node. The number of interconnects is a function of the number of nodes.

The hierarchical topology requires the highest number of non-terminal device nodes. However, it also offers the lowest number of interconnects on each terminal device. This makes the terminal devices easy to manage and operate. The number of non-terminal device nodes can be reduced by increasing the degree of concentration on each node. Full mesh topology has the lowest number of non-terminal nodes, but all devices have a large number of interconnects they must serve. The hierarchical topology with a partial mesh overlay is a compromise (see Figure 2.3). In that topology, there are fewer ports per terminal device than in the full mesh scenario, and fewer interconnects to serve, but there are still hierarchical paths and direct interconnects between nodes and high-traffic node terminal devices. As a result, these nodes and terminal devices are manageable but reduce the need for a hierarchy of all traffic to the directly interconnected devices.

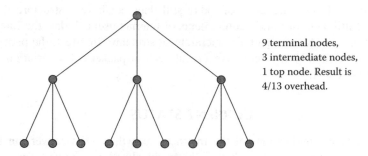

9 terminal nodes,
3 intermediate nodes,
1 top node. Result is
4/13 overhead.

FIGURE 2.1 Hierarchical network with three layers.

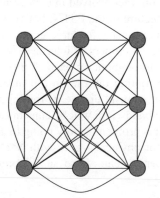

FIGURE 2.2 Nine-node full mesh.

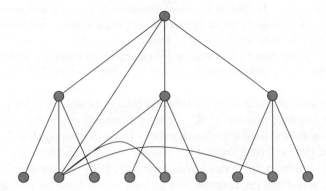

9 terminal nodes,
3 intermediate nodes,
1 top node. Result is
4/13 overhead.
Direct connections
between heavy traffic
nodes.

FIGURE 2.3 Hierarchical network with partial mesh overlay.

Reliability, availability, and serviceability suddenly become important for the routers and the Domain Name Service (DNS) servers to resolve names to IP addresses. Since virtual connections are now set up, tunnels are set up, bandwidth assigned and reserved with the setup and rerouting on a per-packet basis is not feasible anymore. The routers themselves must be built such that they have redundancy provisions. Whereas in the early days of the Internet, every packet could find its way from the source to the destination on a hop-by-hop basis and every packet could indeed take a different path, this is not true any more. With "Quality of Service" (QoS) requirements for the Service Level Agreements (SLAs), latency and delay must be not only closely monitored, they must be controlled and maintained. Rerouting a packet therefore is not a solution anymore, and the entire path to which a transmittal of data has been assigned must be maintained for its duration. This means that the routers must provide redundancy.

This directly affects the line card with line-specific logic and the higher layer processing, as well as the switch fabric cards.

Additionally, Advanced Intelligent Network (AIN) services and Domain Name Service (DNS) lookup require a much more centralized IP signaling network than ever conceived before. The cost of running a DNS lookup engine or center and keeping it current is not negligible. Therefore, network nodes will grow bigger and bigger, and the DNS root servers will also grow in size and complexity. This means that more and more traffic will pass through a decreasing number of very heavily utilized nodes. These nodes will make the Internet increasingly vulnerable and be in absolute opposition to the original idea of the distributed and fail-safe Internet. In essence, the Internet will look like the Telephony Network today, with multiple layers of aggregation hierarchy, local switching and routing, and global switching and routing. DNS lookups will be very much similar to what today is the Intelligent Network (IN) or Advanced Intelligent Network (AIN) lookup in the Public Switched Telephony Network (PSTN) Plain Old Telephony Service (POTS) Time Division Multiplex (TDM) world for 800 numbers, as well as for number portability; Route Planning and Traffic Engineering will make the Internet look a lot more like today's POTS PSTN than the current Internet. We can see first indications of this in Internet2.

DoS and DDoS (Distributed Denial of Service) attacks will be more and more common, and therefore, all crucial network resources will require more and more protection. The Root DNS Servers will have to be faster, will enable more lookups per second, will require a protocol that is secure by means of authenticity verification and data integrity, and encrypt signaling metadata. The routers that communicate with these Root DNS Servers must be able to support this. Therefore, not only the OAM&P cards in future IPv6 routers will have to be able to communicate using the Public Key Infrastructure (PKI), but also Border Gateway Protocol (BGP), Exterior Gateway Protocol (EGP), Classless Inter-Domain Routing (CIDR), or Routing Information Protocol (RIP) might have to use it. They will have to do this within the network, and from the router towards the Root DNS Lookup Center. Messages, commands, and status information will be sent out-of-band as well as in-band, and Service Level Agreements will dominate not only the contracts between end users and carriers or ISPs, they will be prevalent in inter-carrier and inter-ISP contracts. Because of these and because of unification of traffic carried through the backbones, SLAs and QoS will become commonplace.

TRAFFIC AND TRAFFIC GROWTH

The Internet has grown quite dramatically since its inception. Although the predicted traffic growth numbers presented by Vint Cerf never really materialized, the growth of the Internet and the traffic that traverses the Internet on a daily basis is tremendous. Just to put a few throughput numbers into perspective, if every person living in the US is using the Internet via a modem and consumes the available baudrate of 56 Kbits/s, then the grand total bandwidth consumption for all of the US is 300M * 56 Kbps = $16.8 * 10^{12}$ bit/s, or around 17 Tbit/s. To up the numbers slightly, I will assume that every person is using the Internet via ADSL and consumes the available downlink baudrate of 1500 Kbit/s, then the grand total bandwidth for all of the US is 300M * 1500 Kbps = $450 * 10^{12}$ bit/s, or 450 Tbit/s. Even if everyone is online all the time via Symmetric High Speed DSL (G.SHDSL) with symmetric 44 Mbit/s, then the total bandwidth consumption is only 300M * 44000 Kbit/s = $13.2 * 10^{15}$ bit/s, or 13.2 Pbit/s.

Realistically, if the service penetration is lower than 100%, and not everyone is using the Internet, and even if they are, they do not use it 24 hours a day. As a result, we have to significantly decrease the upper limit for expected traffic. If service penetration is 10%, and 1% of the time of the day is spent using the Internet, then we have to take 1/1000 of the numbers we calculated above to get a realistic current traffic upper limit.

As a result, we can assume the upper limit of the bandwidth consumption to be in the range of 10–15 Tbit/s during the highest load period of the day for all of the US. In 1985, a few hundred thousand Kbytes were transferred per day, with peak consumption of a few Mbit/s. Therefore, we can assume that over the last 20 years, traffic increased by a factor of about one million. This is roughly equivalent to doubling the traffic each year over a period of 20 years. Doubling the traffic annually means a Compound Annual Growth Rate (CAGR) of 100%. No technology has withstood a prolonged period of time of hypergrowth.

The basic IPv4 router described was intended to route traffic at speeds between modem lines (then 1200 bit/s), and 10Base2 (10 Mbit/s Carrier Sense Media Access with Collision Detection (CSMA/CD)) at maximum. For the most part, even servers could not load a 10 Mbit/s line, and therefore the requirements were not yet able to sustain a line rate at minimum packet size. The requirement was to support it somewhat, to the best of its abilities—best effort. This is not an option in a 16- or 32-port 10 Gbit/s per port core IPv6 router anymore. As a result, the implementations of the new architectures can be expected to differ significantly.

No matter how good the original design was, after 20 consecutive years of consistent doubling of traffic, the infrastructure of the Internet must be different from how it began.

These numbers are consistent with the starting numbers and the growth rates that MCI published after their "irregularities" were removed, and have roughly been confirmed by Sprint and other carriers that operate data network backbones or the central nodes MAE West and East.

Even if we assume a different distribution of data consumption, and mix services and service requirements as shown in the table below, the total consumption of bandwidth is significantly lower than many experts had anticipated. Even if High Quality or High Definition Television (HDTV) is added to the mix, the total required bandwidth is in the range of 4 Terabit/s for all of the US combined. With routing inefficiencies of a factor of 10—which accounts for signaling and, more importantly, for hierarchical traffic—the required bandwidth grows to 40 Terabit/s. Even if only routers with an aggregate throughput of 1 Gigabit/s were considered, then 4,000 of those would fulfill the needs. (See Table 2.1.)

Additional bandwidth-consuming services will be emerging very soon, and occupy dramatic bandwidth throughout the network. Fleet management in railways

TABLE 2.1
Hypothetical Total U.S. Bandwidth Requirements

	Traffic Volume Estimate (in bit/s)			
Traffic Type	Bandwidth (in bit/s)	Penetration	Usage (sec/day)	Total Data Volume
POTS	9600	0.90	3600	9.331E+15
Cell phone	9600	0.50	3600	5.184E+15
PC & modem	56000	0.50	3600	3.024E+16
PC & ADSL modem	1500000	0.15	3600	2.430E+17
MP3 or similar	10000	0.10	3600	1.080E+15
MPEG or JPEG	200000	0.05	3600	1.080E+16
HQ video	1000000	0.01	3600	1.080E+16
Total data volume	3.104E+17			
Bandwidth required	3.593E+12			
Total U.S. residents	300,000,000			

and trucking companies will very soon rely upon Global Positioning System (GPS), and all the feedback, route info, directions, and map updates will be sent throughout the network, go into Mobile Switching Centers (MSCs), and from there into the cars, trucks, and other vehicles. This may happen using GSM/PCS (Global System for Mobile Communication/Personal Communication Service), but it could also take place using Generalized Packet Radio Service (GPRS), Universal Mobile Telecommunications System (UMTS), Wireless Application Protocol (WAP) and High-Speed Circuit-Switched Data (HSCSD) or other new upcoming technology. Independent of the particular implementation, it will not only occupy bandwidth, but will be of a certain structure. It will have associated signaling like Q.931/E.164, it will use small packets or will be cell based, and it may carry IP traffic. Nevertheless, it will send much more signaling information than current technologies, and it will ubiquitous. Other wireless applications include, but are not limited to, Bluetooth, 802.11a/b/g, Digital Enhanced Cordless Telecommunications (DECT), HyperLAN II, and HomeRF, with upcoming home networking applications like Home Phoneline Networking Alliance (HomePNA)—all with their own inherent signaling and messaging mechanism, and all of them requiring a virtual overlay network for signaling and messaging. Additionally, once these devices are wired, they will be accessed from everywhere, especially from abroad. Roaming information, together with the data to be transferred, will accumulate into quite a large portion of total traffic.

Cell phone usage will continue to increase, but probably not at the rate predicted in previous years. The cell phones will become smarter, and the networks will become smarter in order to provide the user more intelligent services. Experts assume that the traffic generated by cell phone users grows faster than expected, despite the fact that fewer new subscribers than projected will be using wireless services. What is important to mention here is the fact that, especially in the Scandinavian countries, micropayment methods with cell phones have gained quite a bit of ground. People try not to carry cash around any more, and credit card purchases for small payments do not make sense because the administrative effort and therefore the cost for that is too high. Therefore, micropayments are carried out using a cell phone, and the amount due is billed to the phone bill. This increases the amount of cell phone usage with very short messages, close to Short Message Service (SMS) usage. Any new infrastructure in the network must be able to efficiently route and switch these messages at an appropriate granularity.

Additionally, there will be increasingly intelligent gadgets using the cell phone infrastructure, but which are not cell phones. Automotive anti-theft systems and cars equipped with systems—those that automatically dial emergency numbers such as 911 if the car's electronic systems detect a deceleration exceeding a certain threshold, typically indicating a severe accident, and other systems to locate and triangulate those cars—will generate additional traffic.

Some other trends may also contribute to additional traffic on the network. Experts envision that the open network will be used more and more for business and mission-critical transactions, and therefore authenticity and security will become more and more of an issue. This induces the need for trust centers, which in turn will have to authenticate users and validate keys. This requires reliable and high-priority links into the trust centers, to be used for each secured transaction.

This generates additional traffic, but in contrast to other traffic, is extremely important and vital to the users. This will replace Virtual Private Networks (VPNs) entirely. B2B and B2C will be unthinkable in the future without trusted communication over the open Internet. Trust centers will be a crucial part of it, and they will generate a significant amount of high-quality traffic.

In addition to these, the increasing number of Top Level Domains (TLDs) and generic Top Level Domains (gTLDs) as well as hosts at all will lead to a much higher rate at which address resolution requests cannot be fulfilled locally and therefore must be resolved at a very few central places—much like the Intelligent Network with the 800 numbers. This will lead to additional traffic for address and name resolution services, and may open up a completely new market. Search engines with better address resolution capabilities and RealNames equivalent services will help the user find information, but at the expense of a lot of lookups in the Net: short and timing-critical messages to be exchanged everywhere.

Because of integration, unification and convergence requirements for both the carriers and the data service providers such as ISPs, the Internet is already undergoing dramatic changes. While in the beginning the Internet used the PSTN TelCo infrastructure, the TelCo industry has started to use the Internet infrastructure as a backbone.

This is a very fundamental change, and the change is apparent in the backbone and in the access area. In the early days, Class 5 Central Office (CO) switches were the concentrators for end users with dial-up ports and modems. Class 5 CO switches now provide DSL ports for high-bandwidth users, but the DSL ports are not routed to the Class 5 switch core anymore. Instead, they are physically and logically separated and connected to ATM switches or Digital Subscriber Access Line Multiplexers (DSLAMs). The DSLAM ports now connect to carrier-grade routers instead of Class 4 CO switches. In the backbone, the change is equally dramatic. The Class 5 CO switches were connected to a hierarchy of Class 4 CO switches through T1/E1 or higher multiplexed digital hierarchies (Plesiochronus Digital Hierarchy, PDH). In fact, the Class 4 switches were the backbone of the Internet. The PDH uplinks from the Class 5 switches now connect to carrier-grade routers instead of Class 4 CO switches, and as a result, the Internet is the backbone for the TelCo industry, the carriers, the RBOCs and CLECs, and the datacom industry. It is the core of the integrated, unified, and converged network.

The unified network in the backbone and in the access area will force all providers—the carriers, the RBOCs and CLECs, and the datacom industry such as the ISPs—to focus more on International Telecommunications Union—Telephony Section (ITU-T) standards for reliability, availability, serviceability, and manageability, and therefore demand more from future routers. We can conclude that future IPv6 routers will have to meet the following requirements:

- Losslessly switch and route datagrams
- Be able to perform Segmentation And Reassembly (SAR)
- Perform policing at line speed for thousands of virtual connections simultaneously and thereby enforce and support SLAs

- Perform Traffic Management by queuing and buffering traffic according to SLAs, and drop excess traffic
- Support Traffic Engineering by means of collecting Statistic Traffic Data
- Support a mixture of hierarchical and mesh interconnect infrastructure in terms of data traffic
- Support an overlay network for metadata and signaling data
- Enable traffic rerouting at the edge and provide redundant fail-safe systems towards the core
- Communicate securely within the components of the router
- Communicate securely between the OAM&P card and a Billing Center
- Communicate securely between the OAM&P card and a PKI Center for authentication
- Communicate securely between the OAM&P card and a Network Management Center (NMC)
- Communicate with the PSTN infrastructure.

In addition, they will have to do what the traditional IPv4 basic router always did and must do—route traffic efficiently. However, they must do it at dramatically higher speeds, with more line cards per router, and supporting more protocols simultaneously. This gives us another set of requirements that future advanced routers will have to fulfill.

CONCLUSION

The original architects of the Internet and subsequently the first architects of routers have done a tremendous job. They have created a basic architecture that has, so far, withstood the test of time and survived a 20-year run of annual doubling of traffic. The newest changes in the requirements from the view of network operators, carriers, ISPs and users will result in a new form of the Internet. IPv6 and QoS within Service Level Agreements will become predominant, and the Internet as we can envision it today will replace the PSTN backbone. As a result, the Internet and modern routers will have to support new requirements.

3 The Carrier Business Model

OVERVIEW

Data services—especially IP services—always held a promising future that has not yet come to economic fruition in the worlds of the carriers and ISPs. Neither has Voice over IP (VoIP) displaced Plain Old Telephony Services (POTS), nor have the ISPs and the data carriers had great financial success so far.

One of the many reasons IP services have not yet really taken off is the fact that the promoters of IP services ignored one important prerequisite, and that was and is that the Service Providers—carriers and ISPs—must make a profit. This requires a business model that includes billing, and billing requires demonstrable compliance with SLAs and Quality of Service (QoS) promises. Although the sheer volume and the associated cost of IP traffic suggest that every packet is free in terms of expenses, it really is not. The infrastructure must be in place, and it must be paid for; even IP traffic has to be billed somehow. Comparing this to "Good Old Telephony" the per-packet costs of data transmitted over an IP network indeed is cheaper than the "per timeslot" cost of each and every phone call. However, a survey within the peer groups of Telephony Service Providers showed that up to 80% of the voice call costs incurred are costs for billing. Nobody is sure whether these numbers still hold true, but this should teach the router manufacturers an important lesson: if the router does not support accounting for traffic and therefore billing, then the total cost of the transmission of any datagram may be lower; but since none of the traffic is billable, the service providers cannot afford to buy them since they cannot afford to deploy it. They must make money by billing. Flat-rate billing is unfair, unjustifiable, and probably does not hold its position in court. Why should a person or an organization that transmits 100 Mbyte of low-priority traffic per month pay the same flat fee as a person or an organization that makes use of much more bandwidth? While the infrastructure cost surely requires a basic fee for making services available, every bit counts (and must be paid for). High-priority traffic is more expensive, so a person or an organization that consumes 1 Gbyte of high-priority bandwidth occupies many more resources than the above-mentioned standard user. So, at least for internal billing purposes, per-priority billing and policing must be an option. As long as this is not the case, router manufacturers will not sell too many routers into TelCo and backbone data service providers.

Billing is only then a possible business suggestion if and when there are SLAs and a set of QoS parameters in place. In turn, this requires billing on a per-cell, per-packet, per-connection, or per-transaction basis. This requires the router to be capable of not only supporting full QoS and enforcing SLAs, but policing on a per-cell basis and sending this information to a billing center. This transmission must be sufficiently secure that it cannot be cracked or its integrity damaged by any reasonable means. Network Management and IP Billing solutions are necessary for IP Services to be a global success—and IPv6 is crucial. However, this will put additional stress and burdens on the router, since it now must be able to police every single packet. It must perform statistical traffic analysis and monitoring, and it must be able to monitor traffic on a per-user level; if that user has subscribed to multiple QoS levels, the router must track all paths, connections, transactions, and quotas for this user. It additionally must update usage tables, thresholds, billing information records, and all information pertaining thereto. After each transaction or after a preset time interval, the router will have to send this data to the billing center within the Network Management Center (NMC).

Today, the requirements are mostly determined by SLAs between the Service Provider and the customer. In ATM networks, there are multiple priority levels to allow a customer to choose the right combination of services for his demand and particular situation. Mission-critical data—like Session Announcement Protocol (SAP)—will not easily tolerate dropped packets, and therefore "best effort" might not be a good choice. Real-time requirements for video distribution can be fulfilled at a price. For NNTP, Simple Mail Transfer Protocol (SMPT), File Transfer Protocol (FTP), and TFTP, "best effort" is good enough. If VoIP is used, a TDM-equivalent clear channel might be useful. However, all these services must be billed differently. In turn, this requires that the router is capable of associating paths or transactions to a user, combining that with the QoS and the billing information on record, and then determining whether or not the user is within his or her quota. If so, everything can proceed, but if not, a notification to the billing center or the OAM&P center within the NMC must be sent and confirmed.

Very soon the CLECs, the RBOCs, and the long-distance telephone operators and providers will understand that the pure price pressure will force them to abandon their telephony Class 4 switches, and consolidate Time Division Multiplex (TDM) traffic originating and terminating in Class 5 switches into Gigabit- and Terabit-switched routers (GSRs and TSRs). Consequently these devices will have to be able to route and interpret ITU-T Common Channel Signaling System number 7 (CCS#7) messages. Even if Voice over IP (VoIP) will not penetrate the access quickly, it will be used in the backbone. The long-distance telephony traffic will ultimately share the bandwidth in the backbone with data. So there will be timing- and delay-variation-sensitive phone traffic and timing-insensitive data traffic in the backbones; for cost and Operation, Administration, Maintenance and Provisioning (OAM&P) reasons, however, the Class 5 switches will remain for some time. While the impact on the long-distance providers and carriers is quite profound, the cell phone providers and carriers will feel the pressure even more. Not only have cell phones started to displace landlines, the pricing has come down to a level that is on par with the landline—at a higher level of capital expenditure than for wirebound phone services. In order to survive that crunch, cell phone providers and carriers must reduce their operating cost below that of a wireline phone provider. That is only possible for the cell phone operators if they effectively can transport the data and the signaling, including all metadata, for free. As a result, they must replace all Class 4 and Class 5 CO switches with a router-based network and maintain gateways to the PSTN and only keep their "air interface," the Mobile Switching Centers (MSCs).

A QoS-aware and SLA-capable network infrastructure must be in place to make IPv6—or any other advanced protocol, for that matter—a viable alternative to TDM backbones; therefore routers must support these functions in hardware and software for the anticipated number of simultaneous users.

SERVICE LEVEL AGREEMENTS (SLAS)

IPv6 itself is a quite dramatic technology change, but the technology change in the protocol is eclipsed by its implications. In fact, IPv6 brings the Internet Protocol and all services based upon it closer to ATM than most proponents want to admit. Since IPv6 is capable of unifying data and voice services—a promise that ATM made but never really fulfilled—both types of services will have to be supported by IPv6.

That is exactly what IPv6 is all about. IPv6 defines a Quality of Service field that ultimately can be used to implement all the traffic types that are in use today. Where ATM defined Constant Bit Rate (CBR), Variable Bit Rate/Real Time (VBR-RT), Variable Bit Rate/Non-Real Time (VBR-NRT), Available Bit Rate (ABR), and Unspecified Bit Rate (UBR), IPv6 defines 16 priorities that can be used to implement several different levels of QoS. However, IPv6 is not as sophisticated as ATM because it does not have to be. Where ATM defined all these parameters for the variance of the bit rate and the cell delay variation (CDV), IPv6 defines only levels of priority for QoS, and that is ultimately what the routers and their line cards and switch fabrics can understand and process. While it is nice to have CBR, VBR-RT and VBR-NRT available, it makes the design of a line card extremely complicated because arrival and departure of cells must be policed and monitored. It is computationally much more useful to reduce the complexity down to priorities with which cells are processed and forwarded from the ingress line card to the egress line card through the switch fabric. In effect, IPv6 provides the same or better throughput at the same or better compliance with prioritization as ATM, but it does not provide guarantees for real-time transmittals of data. There is no provision for measuring or influencing Cell Delay Variation (CDV) or packet delay variation in IPv6.

For all traffic types that are included in the SLA, important criteria and parameters for SLAs will include:

- Net bit rate (minimum, maximum, average)
- System availability (minimum)
- System uptime (minimum)
- Cell Delay Variation (minimum, maximum, average) (ATM networks only)
- Logical Connection setup time (minimum, maximum, average)
- Delay and latency (minimum, maximum, average)
- Round trip delay (minimum, maximum, average)

As a result of these SLAs, both the service provider and the customer will have to have equipment in place that can validate or verify compliance or noncompliance with the agreed-upon parameters. Another consequence is that the service providers will have to put equipment in place that has at least a chance of complying with the SLAs' parameters.

Therefore, router manufacturers must design a new generation of routers capable of fulfilling the above-mentioned requirements. In other words, the carrier requirements will have to be translated into requirements for the advanced router architectures.

THE VoIP BUSINESS MODEL

Independent of whether the carriers, long-distance providers, or the ISPs will start billing VoIP calls (including higher bandwidth video-telephony calls), there must be a mechanism in place to associate an SLA with a call or a similar customer transmittal. Billing of telephony and data services today is based on a general contractual framework. The call itself has implicit implications. The call does not establish a contract with an SLA or a QoS. In the future, the contract—including SLA and QoS—can be made or modified and agreed upon by the service provider and the customer prior to any call. A flat rate would be an alternative, but then there is no possibility of modifying the QoS parameters as required.

Today, based upon the contractual framework, every single call is carried by the provider, assuming the client agrees upon the conditions that are currently valid at that point of time. This implies uncertainty for both the customer and the service provider. While the customer has no proof for the beginning of, the length of, and the route taken by the call, the service provider has no assurance that the person who initiated the call was legally empowered to do so. This leads to the fact that the customer typically has no certainty about the cost of the call while the service provider has no certainty if and whether the invoice for the call might be disputed. The service provider may have deployed Least Cost Routing, but the customer has no direct benefit out of it. The customer therefore cannot adjust his habits to periods of time in which the service providers' resources are available, or at least when the likelihood of availability of delay-bound bandwidth is higher. On the other hand, the service provider has no impact on the habits of the customer, so overprovisioning of bandwidth is necessary. This leads to higher cost.

It therefore would be appropriate if the resources could be used in a more efficient way. This could be accomplished by providing for the service providers' network a resource manager that is capable of negotiating every single call over the network. The network in this definition can comprise of not only the telephony network, but also of any other resource capable of transporting, switching, routing, or bridging the requested traffic type in a mutually agreeable manner from the source to the destination—including provisions for buffering the traffic, delivering well-defined Classes and Qualities of Service, cell and packet loss probabilities, cell and packet delay variations, and cell and packet reordering.

If the service provider and the customer agree upon this scheme, then it is a natural implication to negotiate a contract for every single call and therefrom mutually determine the quality and the price. Since the resources have a certain value at any given (but variable) point of time—depending on network utilization and load status—the fees or charges cannot be constant. They might even change during the call or the data transfer. This implies something else: the framework in between the customer and the service provider covers only very basic responsibilities and duties, as well as the charges for these. Every single call or data transfer request establishes a new, explicit contract. The customer requests a certain connection for a given time or requests a certain bandwidth along a path with well-defined requirements. The resource manager figures out the actual value of this connection and routes it on the shortest, most cost-effective way or any available path from the source to the destination. The caller then has better control over his willingness to pay for premium quality or to save money by using spare bandwidth, or even to decide to postpone the call or data transfer; the provider has the opportunity to better load his valuable resources. There are no more implicit contracts, no fixed communications schemes, and no unnecessary blocking of valuable resources.

This is a much more demand-driven approach, an open market: it provides the freedom and the fair negotiation between a requester and a resource provider. Valuable resources rise in cost, while spare bandwidth and resources decline in cost. This generates a much smoother usage pattern, better overall network load and utilization, and therefore a higher return of investment for the provider, since its effort to extend the network is now reduced. However, this is only possible with a dedicated and sophisticated capability for policing, billing, traffic management and engineering, and OAM&P support.

INTERNAL CORPORATE ACCOUNTING

While billing for IP services by carriers or ISPs may not be immediately imminent, another topic is apparent today: internal accounting for services in corporations. Therefore, billing through the ISPs and carriers is not painting the entire picture. All Information Technology (IT) must increasingly justify its expenditures—and rightfully so. In all organizations that hold their IT department accountable, accountability is necessary, and it is crucial. All IT spending must be justified, and that is not easy today. While it is possible to say peak load of some routers may be 80% of their capacity, that does not say anything about usage and consumers of bandwidth. Today, at best, a Chief Information Officer (CIO) can claim that he assumes from some probes or statistical monitoring of the network that there is a certain distribution of bandwidth consumption. This is impossible to track over time or on a per-user, per-priority class. In other words, there is no clear technical justification for more bandwidth. What would it be like if the CIO could report to the Chief Financial Officer (CFO) that the internal network must be upgraded because engineering requires 34% of all traffic, with 60% of priority 4 traffic, 15% of priority 3 traffic, 15% of priority 2 traffic and 10% of priority 1 traffic; internal business reengineering through SAP consumes 33% of all traffic, with 5% of priority 4 traffic, 10% of priority 3 traffic, 30% of priority 2 traffic and 60% of priority 1 traffic; revenue-generating orders from outside take 33% of all traffic, with 10% of priority 4 traffic, 5% of priority 3 traffic, 5% of priority 2 traffic and 80% of priority 1 traffic? That would be a perfect justification to order not only more routers to extend the capacity of the network, but also to adjust SAP traffic in its volume and priority.

This would allow for internal accounting of bandwidth usage, and would enable the cost of the infrastructure to be shared and distributed in a fair way—according to its consumption. This would remove the dark cloud over IT infrastructure investments, which today are seen as necessary. They are accounted for by a non-transparent account and seen as an expense for which justification cannot be obtained—other than the statement of the CIO saying "we need it to function."

As a result, the expense for the network infrastructure can now be billed directly towards the departments' accounts, instead of against a general corporate or company-wide "Network Infrastructure" account to which all departments contributed, independent of their usage profile. This may also curb at least some appetite for additional bandwidth, and it might also help pinpoint individual bandwidth hogs. The reason for that is the tendency to waste whatever is free, but be thoughtful of whatever has a cost associated with it.

CONCLUSION

Carriers and ISPs have offered data services quite successfully for a while. As it became increasingly obvious that the data network was perfectly capable of carrying even traffic that had some real-time requirements, this type of traffic increased. Ultimately, it showed that IPv4 and "best-effort" traffic routing could not deal with the complexities of real-time traffic and different priorities. As a result, IPv6 was called to duty for multiservice traffic to unify the carrier's backbones and access network and to carry the traffic that stems out of the PSTN backbone. While billing and policing were not in the original blueprint of the current data networks, these features can easily be supported and implemented by the next generation of smart routers. In order to fulfill the requirements of the new carriers, ISPs, corporate backbones, and the access networks, they will have to:

- Losslessly switch and route datagrams
- Be able to perform Segmentation And Reassembly (SAR)
- Perform policing at line speed for thousands of virtual connections simultaneously and thereby enforce and support SLAs
- Perform Traffic Management by queuing and buffering traffic according to SLAs, and drop excess traffic
- Support Traffic Engineering by means of collecting Statistic Traffic Data
- Support a mixture of hierarchical and mesh interconnect infrastructure in terms of data traffic
- Support an overlay network for metadata and signaling data
- Enable traffic rerouting at the edge and provide redundant fail-safe systems towards the core
- Communicate securely within the components of the router
- Communicate securely between the OAM&P card and a Billing Center
- Communicate securely between the OAM&P card and a PKI Center for authentication
- Communicate securely between the OAM&P card and a Network Management Center (NMC)
- Communicate with the PSTN infrastructure.

We will show in the following chapters what impact these requirements have on advanced router architectures.

4 Advanced Routers in Central Office Applications

OVERVIEW

Many advanced routers will be set up in Central Office (CO) locations. The reason for this is that routers will replace many Public Switched Telephony Network (PSTN) Class 4 CO switches. Bandwidth in the core network will become more important to carriers. The shift of paradigm towards an IP backbone has already taken place. While the Class 5 CO switch will remain in place for a while, the backbone will be entirely IPv6-based, and as a result, the routers in CO applications will have to take over functions of the Class 4 switch. Consequently, the carriers will make sure that routers replacing the Class 4 switches will fulfill the same environmental and space requirements.

CENTRAL OFFICE (CO) COLOCATION

In a Central Office (CO) colocation scenario, there are requirements for maximum heat dissipation, air-cooling and airflow, as well as temperature gradient definitions, maximum power consumption, and resistance to mechanical vibrations of any device that is located in the CO. Colocation is essential for most datacom devices that are used in a carrier environment, such as IPv6 routers. By colocating equipment with the CO equipment, not only is there direct access to all transmission equipment (DWDM, SDH/SONET), but also access to Intelligent Network Lookup servers, NMC alert systems, and—last, but not least—battery backup and in some cases, diesel backup generators. As a result, the router will experience blackouts (loss of power) and brownouts (line voltage reduction) very infrequently, and therefore will very rarely be required to perform a planned shutdown because of a lack of power. However, it will have to prove that its reliability and architecture supports a 99.9996% or better system availability, translating to an unplanned downtime of 3 minutes or less per year. The routers will have to be rack-mounted, and they likely will have to use 23″ racks, since that is the usual standard in TelCo systems. Support of the cabling infrastructure and accessibility of the modules for servicing become important items. Since an IPv6 router will likely be used in mission-critical applications and environments, even the chassis design is different from typical IPv4 routers. It must be designed such that it can run on –48V DC power supplies or 220V AC supplies. This is necessary to feed enough power into the router. Most likely, a carrier-grade IPv6 router will have a 3000W or higher power consumption due to its required throughput, the number of ports, and the amount of processing required. 3000W will require a 220V AC power supply or a –48V DC supply, which is the standard in PSTN CO battery backup systems. At 110V AC and 16A per circuit, the maximum power supplied into the router is 1760W.

Since these routers very likely will be in a Network Equipment Building Standard (NEBS)-compliant environment, they will have to support convection cooling together with forced cooling. In other words, they will be placed in a rack bolted to a raised floor supplying cooled air from the bottom and dissipating heat into the suspended ceiling, which in turn directs the heated air back to the air conditioning. The racks themselves and all devices within them must be designed such that no large horizontal objects obstruct the convection airflow.

The racks most likely contain fan trays that force cooled air from the bottom through the raised floor up to the suspended ceiling. Therefore, the routers must not contain any components that will significantly obstruct that airflow. They must have all Printed Circuit Boards (PCBs) arranged vertically and provide enough ventilation through the chassis to cool the internal components without significantly blocking the airflow through the rack. In some cases of very high power consumption, these routers will have to provide their own fan trays to support rack ventilation. As a result, single-board routers with one large horizontal mainboard will not be allowed in CO locations.

In some cases the Central Office parameters and requirements are such that a router must survive, for a few minutes, an outage of forced cooling or a complete shutdown of the fans accelerating the airflow. This will require that components in the router itself either withstand these reduced airflow periods with airflow through convection only, and no forced flow, or they must be able to shut down individual components if they exceed a certain preset temperature threshold. In any case, it is important that convection is taken into account for cooling purposes. This will also affect the power supplies or power converters. These likely will be in sets of three, and they will have to be laid out such that they are load-sharing, but can supply the router fully using any two of them while the third is shutdown or removed for servicing.

Cooling a large Central Office is indeed one of the major problems faced by the carrier. In a large CO, a huge number of devices is installed, and all of them dissipate heat. This heat must be removed, since the operating parameters for all equipment are essentially the same. While the electricity required to run the devices is large, the bigger problem looms in the cooling thereof. The efficiency of electric cooling is less than 30%. However, the efficiency to a large degree depends on airflow. The more restricted the airflow, the more hot spots remain, and the larger the temperature difference must be between the hot spots and the ambient temperature to maintain an operating environment that is within the specifications of the device manufacturer. As a result, the router's chassis and rack design directly impact the cooling cost in a CO.

To maintain the temperature in a closed room, 1W of heat generated therefore requires at least 3W for cooling it down. This means that any increase in power consumption by 1W will imply a total increase of at least 4W power consumed at the entire CO. This does not sound like a lot, but the large number of devices with their combined power consumption makes that a large challenge: one typical carrier-grade router consumes 3500W (220V nominal at 16A). There are six routers to a rack if the router manufacturer performed a sensible design. In a typical CO, the colocation for datacom equipment and backbone routers reserves space for either a very few (one to four) or up to 30 racks. This is a minimum of six routers at 3500W, or a maximum of $6*30*3500W = 180*3500W = 630000W = 630$ KW.

The routers in this example dissipate 630KW of electric energy. The power company must deliver at least 630KW for the routers, and another 1890KW for the AC to cool them down into this CO. In total, $2520KW = 2.52$ MW of electrical power must be supplied into this sample CO for the routers and their cooling system—not even counting the rest of the TelCo equipment or the loss at the Uninterruptible Power Supply (UPS) and its batteries.

To understand the magnitude of this power requirement, make a comparison to the power grid. A nuclear power plant typically delivers 900 to 1200MW. For this particular Central Office alone, the carrier consumes 2.5% of the power output of a nuclear power plant for a communications site. It takes one whole nuclear power plant to power just 40 communications sites. This should make it apparent that it pays off to make attempts to reduce the power consumption of routers. Every 1W not dissipated saves at least 4W total power consumption.

Even in the US, where electric power is cheap, 1KWh costs roughly $0.11. A year has 8,760 hours. 1W of power consumption reduction saves 8,760Wh annually directly, or $0.96. Combined with the savings in cooling, this amounts to $3.85 annual savings for the operator of the CO. Even if the lifespan of a router in a CO is only three years—which it is not, it is much longer than that—the savings over the life of the router amount to $11.56 for just 1W reduction of its power consumption. It absolutely pays off for the router manufacturer and the customer to make sure that power management is included in the router, and that power-efficient components and architectures are deployed.

If the cost to the router manufacturer is an additional $1 for the hardware or the power-efficient components, $2 for the software to perform power management, and it can sell this for an additional $5 per 1W saved, then the CO operator still saves roughly $6.56 per 1W saved over the life of the component and the router manufacturer still makes an additional $2 per 1W saved per router. In other words, power-efficient architectures can quickly become a selling argument. A useful metric for the power consumption could be the number of switched and processed Gbit/s per W consumed, at a certain level of Quality of Service and availability.

Another very important issue in CO applications is cable management. This sounds like a minor item at first, but it has implications. In CO applications, it is mandatory and important that modules and subsystems can be serviced. At the same time, it is very undesirable to remove cabling for serviceability, and routers that are intended to be used in such applications will have to comply with these requirements. The cabling infrastructure in CO applications can become so extensive that software is required to manage the cables and the interconnects they make. This is especially true in large applications such as in CO colocation situations at carriers, ISPs, Points of Presence (PoPs), Network Access Points (NAPs), or Central Internet Exchanges

(CIXs). The cabling infrastructure requires overview and planning as well as maintenance. As a result, servicing modules or subsystems of any device in a CO should have the capability to be serviced without removing the cables. The obvious conclusion is that cables, and therefore connectors, must be at the rear—versus connectors at the front—of a device. It also means that the cables are not attached to the modules or subsystems, but to the back of the chassis, which stays in place. From there, all cables are routed into the rest of the installation. This implies that port, interface, or line cards do not have connectors at their front, but rather route the connectors to the back of the chassis so that they connect. While this requires one more intermediate internal set of connectors, it allows for the cabling to stay intact if a module or subsystem is removed and reinstalled. This favors backplane designs and not midplane designs or single-board designs. Single-board architectures do not offer serviceability of modules because there are none, and midplane architectures have modules and subsystems on both sides of the midplane, and therefore require service access from the back.

Of paramount importance in Central Office applications are serviceability and safety of the service personnel while servicing a module or a subsystem. All modules and subsystems must be easily accessible, and removal and reinsertion must be possible without additional tools other than the levers on the chassis. The defective modules or subsystems must be unmistakably marked by numbering or—better than numbers—by a red LED indicating the defective status at its front panel. Additionally, the modules and subsystems must be capable of hot-plugging—removal and insertion while powered. It is mandatory that a module or a subsystem fulfills safety requirements, such as eye safety on lasers or electric safety for power supplies. No harmful current or radiation must be present while removing a module or a subsystem.

Also, space is limited in CO applications. Every node must provide the highest throughput within the smallest possible volume, not only in its footprint. While typical TelCo racks are 23″ wide instead of 19″ for datacom equipment, volume is limited. This implies that the number of ports, the line rates, and the throughput per volume unit become important items. If a router requires an external multistage switching extension to perform its function or to be aggregated, then it will take up more volume because it will add the aggregation switch, require more cables to be run between the units, and add to the complexity of the installation. A useful metric for the volumetric density could be the number of switched and processed Gbit/s per volume unit, at a certain Quality of Service level and availability. As a result, careful consideration must be exercised when cascading routers (versus a partial mesh) or a larger router. It might make more sense to aggregate into a larger router that provides a higher performance and offers a better density of switching capacity per volume unit.

In Central Office applications, routers will be subject to the NEBS requirements, and as such will have to be able to withstand certain levels of vibration, shocks, humidity, and changes of air pressure, as well as the range of air pressure. That will in almost all cases require a mechanically stable chassis and locking connectors, but not necessarily significant changes to the modules' and subsystems' electronic systems. It will, however, mostly rule out the use of hard disks, CD ROM or DVD drives, or any other electromechanical or electro-opto-mechanical storage subsystem, if it is not mechanically dampened.

CONCLUSION

The unification or convergence of networks towards an integrated access and backbone network relying on IPv6 has far-reaching implications on routers that are deployed in Central Office applications. They will have to comply with NEBS requirements and all other established standards or habitual agreements that carriers have used for decades. Their reliability, availability, and serviceability will have to reach levels that TelCo equipment has provided for decades, and they will have to perform this at higher levels of performance per volume unit, that the old TelCo equipment will never achieve. At the same time, power consumption and heat dissipation must be kept at a minimum.

5 Function Split

OVERVIEW

Modern routers are complex devices. The complexity is present in their hardware and software. Throughput and robustness as well as system availability requirements force router architects and designers to partition the functions as clearly as possible to simplify system design and testing. The function split is sometimes physical, sometimes logical, and sometimes both physical and logical. The function split can be between hardware and software, and between hardware and software modules. It can be such that a certain portion of software on a particular piece of hardware is a logical unit, and other portions of the hardware and software belong to other logical units. The function split pertains to data path functions as much as it does to control path functions.

While it is absolutely essential that the line cards *prepare* the datagrams for being forwarded by the switch fabric, they should not attempt to *perform the scheduling* within the switch fabric. On the other hand, the switch fabric should only *forward* the datagrams that are delivered to its ingress port to the appropriate destination port—and not perform any *processing* of the datagram. If the switch fabric has to *queue* the datagram because of internal or switch fabric port contention, it can do so. However, it must not try to take over *traffic management functions*. The function split between the devices on the line card is equally important. The network processor extracts routing information from the incoming datagram and prepends a tag that the traffic manager and the switch fabric use. It is not the task of the traffic manager to perform investigation into the datagram to determine its priority and real-time requirements. The optional ingress and egress side traffic managers use the tag that was created by the network processor to determine when and if the datagram can be queued, discarded, dropped, or must be forwarded to the switch fabric immediately. The switch fabric uses the same tag to determine the destination port and the priority of the datagram. Reassembly occurs based on the prepended tag as well—independent of whether it is performed in a SAR engine, or as a task within the traffic manager. The drop policies are executed within the traffic manager, based on the tag and the policy and the quota the network processor has provided. The port card contains physical layer transceivers (PHYs) and Media Access Controllers (MACs) as well as framers to extract the payload, which in turn is then transferred to the processor card along with the required information retrieved from the encapsulation and wrappers. However, the processor card does not try to perform the functions of the port card.

A simple way to understand and visualize the individual tasks of the subsystems is to track the path of a packet through the router. The packet will enter the port card. PHY and MAC will remove the unnecessary parts of the packet and forward it to the network processor. The network processor will then determine its further path through the router by determining its Local Connection Identifier (LCI). The LCI consists of the destination port number, the priority, the real-time requirements, its drop priority, the source port number, and the Cell Sequence Number (CSN) of the datagram after segmentation has occurred (see Figure 5.1). The network processor also looks up the policy and the quota to determine the drop priority.

As we can see, different subsystems of the router and the line card use different portions of the LCI. None of these subsystems or components has to use any other information. This information is extracted by the network processors and prepended to the payload as a tag to guarantee that all subsequent units and engines have the same information available to perform their functions; no contradictory information is used for switching, routing, forwarding, queuing, or discarding the datagrams. While the switch fabric only requires the destination port number and the priority, the traffic manager needs to know the destination port number, the priority, the real-time requirements, the drop policy, and the source port number. The traffic manager will enqueue and dequeue based on additional information, such as its queue status and the egress link status. The same is true for the policing engine. It mostly focuses on the drop policy and determines when and if to drop the datagram. The traffic manager will determine its queue status and the status of the egress link in order to figure out if datagrams must be dropped due to persistent overload. The Reassembly engine needs the source port number and the priority as well as the CSN to reassemble a datagram.

This logical function split is often mirrored in the actual physical setup and in the implementation of the software.

If these functions are not strictly separated, two or more devices or software modules may cover corner cases, each one potentially executing conflicting actions upon the datagram. These corner cases become virtually impossible to find, to test, to reproduce and to fix. As a result, there is no option for designers and architects

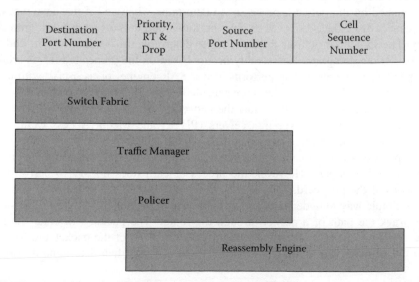

FIGURE 5.1 LCI and its components.

of modern routers. They will have to make sure that each component and subsystem can be tested and verified individually, and then tested and verified as a system. Any ambiguities in hardware or software or the function split thereof will make verification a nearly impossible task.

Ideally, we can separate the functions as shown in Figure 5.2. If we look at the list of requirements that result from the deployment of modern routers in unified networks, we find absolutely crucial the necessity to do the following:

- Losslessly switch and route datagrams
- Be able to perform Segmentation And Reassembly (SAR)
- Perform policing at line speed for thousands of virtual connections simultaneously and thereby enforce and support SLAs
- Perform Traffic Management by queuing and buffering traffic according to SLAs, and drop excess traffic
- Support Traffic Engineering by means of collecting Statistic Traffic Data
- Support a mixture of hierarchical and mesh interconnect infrastructure in terms of data traffic
- Support an overlay network for metadata and signaling data
- Enable traffic rerouting at the edge and provide redundant fail-safe systems towards the core
- Communicate securely within the components of the router
- Communicate securely between the OAM&P card and a Billing Center
- Communicate securely between the OAM&P card and a PKI Center for authentication
- Communicate securely between the OAM&P card and a Network Management Center (NMC)
- Communicate with the PSTN infrastructure.

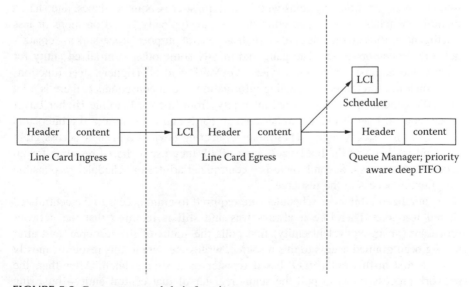

FIGURE 5.2 Components and their functions.

All of the items on the list require functions that are implemented as individual modules—whether in hardware, software, or a combination of both. While in some instances it is useful and appropriate to combine multiple functions into the same piece of hardware, the software must remain modular to allow for easy maintenance of the function, and therefore of the hardware or software module.

Modularity of the hardware and the software is key to the implementation of a complex device such as a modern router. The sheer number of standards and protocols stacked on top of each other and the need for maintenance of these functions requires that there is no ambiguity regarding which module—hardware or software or a combination of both—executes that particular function. Without a clear-cut delineation, neither testing nor maintenance is possible.

Part of the function split is between the routines that process data and control information and the error handlers. They are typically separated. While the modules that process data and control information are implemented as software modules in an outer ring of privilege or priority, the error handlers for software and hardware errors are closer to kernel priority, and therefore in an inner ring of privilege. In some cases, hardware and software errors are handled within Interrupt Service Routines (ISRs). This is a split of functions that makes exception handling significantly easier because the data or control information processing modules only have to deal with foreseeable datagram errors, whereas the ISRs deal with unforeseeable exceptions.

TRADITIONAL SYSTEM PARTITIONING AND FUNCTION SPLIT

Traditional System Partitioning has mostly relied (and still relies) on large monolithic functional units, especially if shared memory switches or buses are used as the interconnecting entities. Systems with buses, rings, crosspoint switches, and shared memory switches as interconnecting devices are typically based on more or less intelligent entities on the line cards—such as general purpose processors as "ersatz"-network processors in the data path, and mostly some other centralized entity for lookup and address resolution services, OAM&P, and all Higher Layer functions (including interpretation of signaling information). As a consequence, there is a lot of traffic generated and terminated internally, from line cards to the Higher Layer processing card and back, and in between all cards and the centralized arbiters and schedulers for access rights. These interconnecting devices and their access grant mechanisms become the bottleneck. First of all, they use a shared resource for all access and transfers. Second, there is a centralized arbiter or scheduler responsible for granting access to the resource.

Centralized arbiters or schedulers are required to grant access to the centralized shared resource. Therefore it always was and still is required that the network processor (or an equivalent entity) first polls the status of the resource, and after having been granted access to this resource, writes the data to this resource, mostly using a first-in-first-out (FIFO)-based register or a simple latch. After that, the network processor has to poll the status register of that central entity—bus, ring arbiter, or crosspoint switch scheduler—in order to find out whether the operation was completed successfully. In other words, there is no clear function split. The network processor (or an equivalent entity) is not only responsible for dealing with the line-specific issues, such as physical layer functions, signaling, protocol interpretation and generation, they are involved in the process of switching the data to the appropriate destination internally as well. Therefore, there is no clear function split in the hardware or in the software. The network processor must take over functions of the central entity to make sure packets or cells get switched, routed, or directed to the appropriate destination; the centralized entity must communicate with the network processor and support and aid in interpretation and translation of the signaling information, therefore ceasing to be independent of the protocol and the signaling. This puts a lot of unnecessary functions into both devices that could easily be avoided by introducing a clear function split.

As a result of the traditional function split, the "ersatz"-network processors performed all functions in a router. Independent of whether there was one processor per port or line card, or one for the entire router, the network processor processed layer 1 to layer 4 information and datagrams including their payload. In addition to that, the network processor controlled the switch fabric, the crosspoint switch, or the shared memory switch, and carried out all OAM&P functions, mostly in a monolithic piece of software.

FUNCTIONS WITHIN THE PORT CARD

The port card is responsible for dealing with layer 1 and 2 processing. It removes and adds encapsulation, frames, and wrappers on layer 1 and 2. It will forward necessary information from the encapsulation together with the payload, if it is required. Also, the MAC will not take over framer tasks, or vice versa. The PHY cannot perform MAC or framer functions, nor can the MAC perform PHY functions or framer tasks. Even if every task of a port card is implemented within one programmable chip, different portions of software on that particular chip will carry out those functions. The port card will not process the datagram or extract information for further processing.

FUNCTIONS WITHIN THE PROCESSOR CARD

The processor card parses the incoming datagram and extracts all necessary information to segment, switch, route, forward, queue, police, reassemble, or discard the datagram. Different engines on the line card perform these functions.

The network processor must perform the lookup of the incoming line-specific address and assign the LCI. This lookup and assignment of the egress port number, along with the appropriate priority for QoS, is essential to the operation of the switch fabric. Since the switch fabric switches cells, it is essential that the Segmentation And Reassembly (SAR) engine on the processor card performs the SAR function. The SAR function can be implemented in software as a separate module on the network processor. Another task on the network processor is the policy lookup for the datagram. Although ingress side traffic management is rare, the ingress side traffic manager, either in hardware or in software as a separate module on the network processor, may do so. However, it must not confuse traffic management with switch fabric queue management. In no case should the processor card perform any switch functions—not even switch control or scheduling, nor switch fabric queue management. The purpose of the switch fabric queuing is to prevent head-of-line blocking and to reduce contention, whereas traffic management on the line cards shapes traffic. This can occur either into the switch fabric, or into the egress port. On egress, the line card's network processor does not have to look up the LCI and policies again because the ingress line card's network processor had already assigned the LCI and the policies. The egress side network processor and the traffic manager are responsible for verifying compliance with policies and quota entries, for billing, for traffic management by shaping, for executing drop policies, and for reassembly of the datagrams. In summary, the line card parses the incoming datagram and extracts all necessary information to segment, switch, route, forward, queue, police, reassemble, or discard the datagram. It prepends the LCI that it determines, based on the incoming datagram's header information to the segments that constitute the internal datagram, so that all subsequent hardware or software modules can perform their tasks without repeating the lookup process. These are very clearly distinctive functions in separate hardware and software building blocks.

FUNCTIONS WITHIN THE SWITCH CARD

Switch cards traditionally were based on shared memory switches, crossbar, or crosspoint switches, without any internal arbiters or schedulers. As a result, some external entity had to take over the scheduling of queuing and switching. Consequently, scheduling the queuing and switching within the switch card had to be performed by the Route and Switch Processor Card or by the line card's processor. A clear function split—logical or physical—was not possible. Newer switch fabrics contain schedulers both for queuing and for crossbar scheduling and control, and as a result, they can switch datagrams entirely autonomously based only on the information in the LCI. The switch fabric uses only the LCI to switch datagrams, and no other external information. It will, however, use the internal crossbar switch status, as well as the status of its own internal queues and links, to determine if it has to enqueue and dequeue datagrams, and when to forward them through the crossbar switch. A modern switch fabric has a variety of different schedulers for switching and queuing. The route and switch processor card is therefore not required anymore to set up the crossbar switch or to schedule the queues. The functions can physically and logically be separated.

FUNCTIONS WITHIN THE OAM&P CARD

The OAM&P entity was traditionally called the Route and Switch Processor Card because it contained the centralized lookup engine and the switch control. In addition to that, it held the centralized routing table and the policy database, and it maintained communication with other routers for route information exchange. As a result, it contained a large portion of the router's software with no apparent partitioning of the tasks. Additionally, it posed a performance bottleneck and a single point of failure. In modern implementations, the OAM&P entity is only responsible for Operation, Administration, Maintenance and Provisioning. It is not performing any switch control, lookup, or routing table updates. It facilitates routing table updates and consolidates statistic traffic data by polling it from the cards in the data path that collect the statistic traffic data. It also consolidates billing information in the same way. This allows for a very clear distinction in its functions, and separates them clearly from other functions in other hardware or software modules.

CONCLUSION

Modern advanced routers are complex devices, consisting of a large number of hardware and software building blocks. A clear function split between the modules makes design and testability possible. More importantly, it allows for maintenance and upgradeability to the hardware and software of the router. As a result, it is paramount that the system architect and designer understand the requirements and their implications on the design, and appropriately partition the functions within the router's hardware and software.

6 High Availability

OVERVIEW

Most routers with advanced architectures will be deployed in the core of corporate networks and in the core of the Internet and the Internet2; a smaller portion of them will be deployed in end customer locations that are not the core or mission-critical parts of the network. While most of the routers in end customer locations will not have to fulfill High Availability (HA) criteria, those in the core of the network—be it a corporate network or the Internet—will. HA is typically defined as fully functional operational status of a device or a network in excess of 99.999%. This is why it is sometimes referred to as "five nines." In other words, the availability of a device or a network is defined in terms of a probability to function as intended, and thus to be available to perform its intended function. Since every device or network consists of a large number of individual components, each of which have their very own life span (and therefore a certain failure probability), the fully functional operational status is possible only when all necessary components function as intended—and that includes their software. As a result, the failure probability of a complex device such as a router is significantly higher than the failure probability of each of its components if no provisions are taken to increase the reliability of the system. As with all compound systems, the failure probability is the compound failure probability of each of the components or subsystems.

DEFINITION

Reliability is defined as the probability that a system, a subsystem, or a module performs its intended function within a specified time interval under the stated conditions.

IMPLICATIONS

Reliability is inherently design-oriented, and therefore its processes and the procedures that comprise the core methodologies are design-related. It becomes immediately apparent that a single system or subsystem will always have a failure probability greater than zero. As a result, failure of a system, subsystem, or module must be taken into account. It is impossible to design and build a completely failsafe system. It is possible, however, to reduce the failure probability to such a degree that it can be considered highly or continuously available. This will by design be a system that is comprised of multiple subsystems, each of which have a certain failure rate. The real secret is to make a system with compound failure probabilities of individual subsystems work more reliably than each of its individual subsystems. There are multiple ways to achieve this.

Reliability is calculated using probabilities. Knowing the probabilities for failure or an outage and the potential impact on the system is therefore crucial to determine which components can be left "Single Points of Failure," and which components, modules, or subsystems must be implemented in redundant pairs or triplets.

Each subsystem must be designed carefully so that it meets its intended availability status. There is no possibility to design a subsystem in a way that its probability to fail is zero, although it can come close to zero. Additionally, there is no need to do so. For example, the panel to display device status is not crucial to the operation of the device itself. Therefore, it does not make sense to spend extra effort to improve its availability from 99.999% to 99.9996%. The backplane or midplane, however, must have a reliability that exceeds by far even the 99.9999% mark.

Therefore, the designer must enable and provide a system design with redundant subsystems, achieving a certain level of tolerance against faults and subsystem or module outages or failures. Fault tolerance can be achieved by providing redundant subsystems. In that case, there is one active subsystem and one hot or cold standby subsystem in each central or partially central unit within a device. For hot standby, both systems are operated in parallel, and both systems process all data in parallel. They may or may not be microcycle synchronous. Typically, lower-speed applications and Telecom systems are microcycle synchronous. In this way, a comparator can compare on a per-cycle basis whether the results of both the active and the hot standby subsystems are identical. If they are, there is certainly no fault in either subsystem because the likelihood of two independent systems producing the same error is extremely low. Cold standby relies on a standby system that is in powered-down mode, and boots up as soon as the currently active subsystem fails and is taken out of order. In an implementation with triple modular redundancy, three identical subsystems feed their results into a device that forwards their result to the next stage if all three inputs are identical; if there is a discrepancy, it forwards the result for which it received the majority of identical results.

In case speed and throughput requirements or the use of High Speed Serial Links (HSSLs) do not allow a microcycle synchronous operation, the system must be set up differently. In that case, all subsystems connected to the redundant subsystems send the data to and receive the data from all redundant units. However, the subsystems connected to the redundant subsystems only forward the data, sent by the active unit, from the transceiver to their internal logic, and discard the data received from the input connected to the standby subsystem. In case the data is sent over HSSL, the transceivers process all layer 1 information and report code violations. Since the data transmitted across two different HSSLs is not microcycle synchronized, it is impossible to perform the switchover completely without loss. At least the current datagram will be lost, if not the packet. To minimize the number of lost datagrams after the switchover decision has been made, a system like this will have to rely on the capability to switch over very quickly after non-recoverable errors have been detected.

Fault tolerance can additionally be achieved by implementing error detection and correction codes in the internal and external communications paths. Additionally, it must be ensured that each and every subsystem and module sends out data according to all pertaining specifications, so that bit errors on the receiving side cannot be attributed to functionally incorrect behavior of the sender. Therefore, if bit errors occur, it can be assumed that they are due to spurious conditions on the internal or external transmission line, and therefore the receiving side can try to recover data using Error Detection and Correction mechanisms. In most cases, checksums are calculated and appended to the datagram or a set of datagrams they span, so that the receiving side can compute the same checksums and compare the datagram or the set, thereof, with its results. Should there be a discrepancy, the error can be detected, and in some cases the error can be corrected. In such a case, it is counted as a spurious error in the registers that will be polled by the OAM&P entity for Statistic Traffic Data Collection purposes, but the datagram or the set thereof is forwarded with the corrected error, and no retransmit request is sent. This means that received data is only discarded if it is unrecoverable or inherently wrong. In this way, even slightly corrupted data can be processed without having to request a retransmit that delays every subsequent datagram. Requesting a retransmit of a datagram certainly is a possibility, but this requires more processing cycles and also delays the reassembly process of the packet—very likely resulting in discarding the entire packet. Recovering the datagram therefore must have priority.

The idea of increasing system availability is to minimize the impact of an error or an outage and to set up the system such that the availability of the system is not based entirely on the availability of any given subsystem, but on the combined reliability of two or more parallel subsystems for all units that directly impact system availability. The system's reliability is based on the compound availability of every single redundant subsystem and the reliability and availability of all devices that constitute a Single Point of Failure related to the redundant subsystem (see Figure 6.1). Good design practices require choosing components for which the reliability and availability of the multiplexer, demultiplexer component, and the select logic is practically 100%, or one—meaning its probability to fail is negligible.

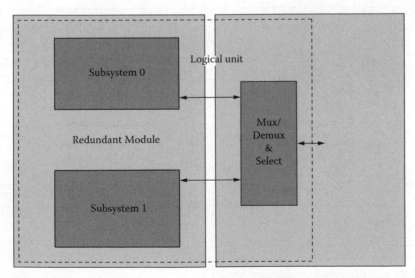

FIGURE 6.1 Redundant subsystem and single point of failure.

While most Single Points of Failure can be avoided, some cannot. Examples for Single Points of Failure that cannot be eliminated are the comparators, the multiplexer and demultiplexer, plus the selection logic for the input selection (see Figure 6.2). As a result, these components must be as simple as possible in order for their reliability to be as high as possible. The intention is to have the failure probability of these components be negligible and therefore not impact the reliability of the redundant subsystem.

It is fairly obvious that the two subsystems should not physically be on the same PCB, and the multiplex and demultiplex chip is a single point of failure. All these components together form a logical unit.

If the probability of a subsystem to fail $p_{F,S}$ is known, then this subsystem is available with a probability $p_{A,S} = (1 - p_{F,S})$; i.e., it is in good working order. If the designer combines two identical subsystems in a redundant setup, i.e., having one active and one hot standby subsystem both working in parallel, then the availability of at least one of the two subsystems is shown in the following equation:

$$p_{A,R} = (1 - p_{F,S}^2)$$

This is because it is only unavailable if both redundant subsystems are nonfunctional at the same time. In addition, the availability of the devices replicating the traffic on the output of the peripheral non-redundant subsystems sender must be factored in, as does the input of this subsystem to select one out of the two inputs it has. Because these devices are typically much less complex than the core subsystems, their failure rate is dramatically lower, typically by orders of magnitude because of the low complexity. Because these parts are single points of failure, the intentional low complexity helps maintain the overall system availability. However, because it is required that the multiplex and demultiplex chip, the select logic and one of the

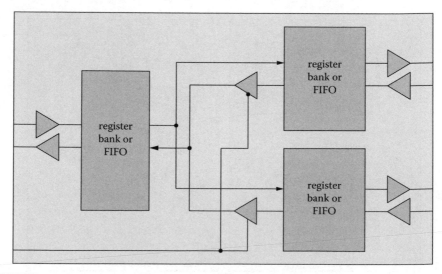

FIGURE 6.2 Redundant transceiver.

two redundant subsystems are in working order. This is a logical AND. In other words, a module consisting of a redundant subsystem with a multiplex and demultiplex chip and select logic is nonfunctional if and only if both of the redundant subsystems, the multiplex and demultiplex chip, or the select logic are nonfunctional.

We must compute the compound probability to fail of the module containing the redundant subsystems with the multiplex and demultiplex chip and the select logic, with $p_{F,M}$ being the failure probability of the multiplex and demultiplex chip and $p_{F,L}$ being the failure probability of the select logic. We can compute the failure probability $p_{F,ML}$ of the multiplex and demultiplex chip or the select logic:

$$p_{F,M} = (p_{F,M} + p_{F,L}) - p_{F,M} * p_{F,L}$$

Then the probability of the entire compound redundant module to fail $p_{F,C}$ is given by this equation:

$$p_{F,C} = (p_{F,S}^2 + p_{F,ML}) - p_{F,S}^2 * p_{F,ML}$$
$$= (p_{F,S}^2 + ((p_{F,M} + p_{F,L}) - p_{F,M} * p_{F,L}))$$
$$- p_{F,S}^2 * ((p_{F,M} + p_{F,L}) - p_{F,M} * p_{F,L})$$

The availability depends on all these components working together seamlessly. The switchover in-between the active and the hot standby should be as fast as possible in order not to lose the entire subsystem due to extended loss of data, and therefore due to the risk of being taken out of service altogether by the OAM&P card. Additionally, the time in which one card is out of service and is not backed by a redundant hot standby must be minimized because nothing prevents an active card failing during the time at which there is no hot standby subsystem in working order. This is one of the reasons why an OAM&P card should notify an NMC immediately, so that the faulty card can be replaced. All events are considered to be independent of each other, i.e., a failing subsystem is impacting neither the availability of the other subsystem, the availability of the multiplex and demultiplex chip, nor that of the select logic. They are statistically independent and stochastically distributed.

High Availability and sometimes even Continuous Availability is required in all Telecom Systems and in a wide variety of networks based on advanced routers. All Class 5 and 4 Central Office switches use redundancy schemes, as do most Internet backbone routers—such as the Cisco GSR 12000/12400 series, the CRS-1, and Juniper's M160 or the T640. Additionally, most of these systems must be able to withstand spurious errors such as injected errors from the outside: corrupted datagrams received; internal errors which can be caused by radiation impact or by spurious and temporary faults such as link errors and permanent faults; or subsystem failure.

NETWORK VIEW OF REDUNDANCY AND FAILSAFE OPERATION

The view in terms of availability has changed such that the network will not require fully redundant and failsafe routers throughout the entire network anymore. To the contrary, in the future fully redundant and failsafe routers will only be required in the core of the network, but the edges—and maybe even the core edge—will rely on path redundancy. In other words, the edge and maybe the core edge will rely on routers that can redirect traffic along redundant paths if the edge suffers an outage. In this way, OAM&P is possible at reasonable cost, and the reliability of the network approaches the required availability for routing even PSTN POTS traffic. So the network-scale QoS and availability guarantee is ensured by inherently reliable devices in the core, and by path redundancies with appropriate OAM&P and NMC involvement in the edge and maybe the core edge. As a result, the network fulfills High Availability criteria, while only core routers must consist of redundant and failsafe modules; the edge and core edge routers just need to provide software on the line cards to provide for path redundancy, in conjunction with the OAM&P and NMC entities.

Redundancy is built into the nodes of the core, not into the entire network anymore. The core routers and switches must be fully redundant to provide reliability and guarantee a certain uptime. Typically, carriers are targeting a maximum of 3 minutes a year downtime of a crucial core network element or node. This number already includes time for software updates. In reality, the device itself is expected to be unavailable for less than these 3 minutes a year from the carriers' standpoint. So if any router or switch is considered for a carrier's backbone, the supplier must make a case for the device. If a typical software upgrade is required once a year, takes two minutes to perform this update, and if the service or availability of the device, then the total remaining downtime for other reasons is one minute.

If every single component is either fully redundant or its function can be assumed by another component—with or without load sharing—then the next question becomes if there is any impact on the Quality of Service or throughput of the node. If that is the case, then very obviously the carrier can only statistically provide the QoS and throughput described and determined in the SLA. This, for the most part, is not acceptable, and therefore most requirements are that even during a failure of a critical part, another identical part can take over. This is true for all routers and switches that are deployed in the core network. On the edge or even the core edge, the outage of a switch or a router will cause the OAM&P entity to alert the NMC, which in turn will immediately make another path available to circumvent the switch or router that failed and dispatch a technician to replace the failed node.

Coming back to the requirements list, we can conclude that future advanced routers—whether IPv6 routers or any others—will have to perform the following:

- Losslessly switch and route datagrams
- Be able to perform Segmentation And Reassembly (SAR)
- Perform policing at line speed for thousands of virtual connections simultaneously and thereby enforce and support SLAs
- Perform Traffic Management by queuing and buffering traffic according to SLAs, and drop excess traffic
- Support Traffic Engineering by means of collecting Statistic Traffic Data
- Support a mixture of hierarchical and mesh interconnect infrastructure in terms of data traffic

- Support an overlay network for metadata and signaling data
- Enable traffic rerouting at the edge and provide redundant fail-safe systems towards the core
- Communicate securely within the components of the router
- Communicate securely between the OAM&P card and a Billing Center
- Communicate securely between the OAM&P card and a PKI Center for authentication
- Communicate securely between the OAM&P card and a Network Management Center (NMC)
- Communicate with the PSTN infrastructure

Again, a direct or indirect impact of these requirements on the router in terms of High Availability is clear. None of the functions can be performed if the system or the subsystem is not available. IPv6 and all other newer and modern protocols on top of IP are themselves quite a dramatic technology change, but the technology change in the protocols is eclipsed by their implications. Since IPv6 and ATM define Quality of Service and are intended to be multi-protocol infrastructures, the goal is to incorporate all PSTN TDM traffic. As a result out of this, system uptime criteria for routers supporting those protocols increased dramatically, especially with POTS Emergency Calls being routed over this infrastructure. But even in corporate environments, the situation starts to change. In earlier times, routers were connected to a Private Branch Exhange (PBX) to connect to the PSTN. The PBX had a "five nines" or "six nines" of availability, and therefore it was virtually always possible to conduct emergency calls. Now that VoIP has gained ground, the router often connects to the DSLAM port of the Class 5 CO switch, and the phones are connected to the router. This means that the emergency call can be placed when and if the router is available, the DSLAM port is accessible, and the lookup engine for IP addresses to E.164 PSTN numbers can resolve the request. As a result, the combined availability of these devices determines the system availability. Consequently, system availability has become a major factor in SLAs.

During the length of the contract, the most important SLA criteria that influence the desired level of availability include:

- Minimum System availability
- Minimum System uptime

These parameters can only be influenced by architecture and design, as well as the implementation of the router.

EXAMPLE

Assume again that p_{FS} is the failure probability of the subsystem, p_{FM} is the failure probability of the multiplex and demultiplex chip, and p_{FL} is the failure probability of the select logic. The failure probability of the compound redundant module is again given by p_{FC}, and its availability therefore is $p_{A,C} = 1 - p_{FC}$.

We can determine the availability of the compound redundant module with the following equations:

$$P_{F,M} = (p_{F,M} + p_{F,L}) - p_{F,M} * p_{F,L}$$

$$P_{F,C} = (p_{F,S}^2 + p_{F,ML}) - p_{F,S}^2 * p_{F,M}P_{F,C}$$
$$= (p_{F,S}^2 + ((p_{F,M} + p_{F,L}) - p_{F,M} * p_{F,L})) - p_{F,S}^2$$
$$* ((p_{F,M} + p_{F,L}) - p_{F,M} * p_{F,L})$$

In order to give a simple example, wrongly assume the data shown here:

$$p_{F,S} = 0.01 = 1 * 10^{-2}$$

$$p_{F,M} = 0.00001 = 1 * 10^{-5}$$

$$p_{F,L} = 0.00001 = 1 * 10^{-5}$$

Therefore, the probability that both redundant subsystems fail at the same time is given in this equation:

$$P_{F,C} = p_{F,S}^2 = 0.01^2 = 1 * 10^{-4}$$

The failure probability for the select logic or the multiplex and demultiplex chip is:

$$p_{F,ML} = (1 * 10^{-5} + 1 * 10^{-5}) - 1 * 10^{-5} * 1 * 10^{-5}$$
$$= (2 * 10^{-5}) - 1 * 10^{-10}$$
$$= (2 - 1 * 10^{-5}) * 10^{-5}$$
$$= 1.99999 * 10^{-5}$$

However, the probability of the entire module being unavailable by means of both redundant subsystems, the select logic, or the multiplex and demultiplex chip being defective is shown as:

$$P_{F,C} = (P_{F,S}^2 + ((P_{F,M} + P_{F,L}) - P_{F,M} * P_{F,L}))$$
$$- P_{F,S}^2 * ((P_{F,M} + P_{F,L}) - P_{F,M} * P_{F,L})$$
$$= (1 * 10^{-2})^2 + 2 * 10^{-5} - 1 * 10^{-10} - (1 * 10^{-2})^2 * (2 * 10^{-5}$$
$$- 1 * 10^{-10})$$
$$= 1 * 10^{-4} + 1.99999 * 10^{-5} - 1 * 10^{-4} * 1.99999 * 10^{-5}$$
$$= 1 * 10^{-4} + 1.99999 * 10^{-5} - 1.99999 * 10^{-9}$$
$$= 0.00011999790$$
$$= 0.011999790\%$$

This failure rate, by definition, is higher than the failure rate of the multiplex and demultiplex chip and the select logic; it is higher than the rate at which both of the redundant subsystems fail simultaneously; however, it is lower than the failure rate of the individual subsystem alone. This is the desired effect. Unfortunately, both the multiplex and demultiplex chip and the select logic are required. Equally unfortunately, the multiplex and demultiplex chip and the select logic pose single points of failure. While the industry has solved the problem of a complex subsystem failure by providing a second hot standby subsystem, there is no way of replicating that for the multiplex and demultiplex chip and the select logic. As a result, designers must make sure that this single point of failure is sufficiently reliable, and that it does not significantly impact the reliability of the compound redundant module. While it would not make sense to increase the reliability of the redundant subsystems to a degree to surpass the reliability of the multiplex and demultiplex chip and the select logic, it makes sense to increase the reliability of all of the said parts, components, and subsystems. However, the increase in reliability must be balanced.

From the above equation, it can be derived that the multiplex and demultiplex chips and the select logic do not significantly impact the availability of the redundant module; the subsystems have a higher impact.

A system with an availability of 99.988000210% (failure rate is 0.011999790%) is statistically unavailable for 63 minutes a year. This is good enough for datacom systems, but not yet good enough for Telecom systems. For these, each the multiplex and demultiplex chip, the select logic, and the independent redundant subsystems must be of a higher reliability. What is most apparent here is the fact that both the multiplex and demultiplex chip and the select logic each contribute the same factor into the equation as does the square of the availability numbers for the redundant components or subsystems. These chips dramatically affect the availability. Therefore, they must be as simple and reliable as possible. To achieve the magic "five nines", (99.999% system availability) these chips and the subsystems must have higher reliability numbers.

If we accept module failures as a given to occur with a certain probability and we provide means for another identical hot standby module to take over, then the question becomes how to maintain system availability during the periods of time in which errors are encountered and detected, the subsystem switchover occurs, and the hot standby subsystem takes over. One obvious conclusion is that all modules must be hot-pluggable. The length of the service interruption then mostly depends on the time it takes to detect a permanent error or outage, the switchover time, and the time to transfer configuration information from the failed module to the replacement module.

The switchover time is determined mostly by the ability to take over of the subsystem previously configured as standby. In some cases, hot standby is required. In a hot standby configuration, both subsystems work in parallel and the switchover command is merely a command issued to units connected to both redundant subsystems to forward one out of two inputs. As a result, the switchover is nearly immediate, and only the "in-flight" datagram will be destroyed, resulting in typically one packet being lost. Cold standby for the most part means that two identical units are physically present, but one is active and the other is either in sleep mode, powered-down mode, or off. In either case, the standby subsystem must reboot or wake up, load all necessary parameters, synchronize the High Speed Serial Links (if present), and then take over the task for the previously active subsystem.

DESIGN LIFE TIME AND SINGLE POINT OF FAILURE IMPACT

The life expectancy (Design Life Time, DLT) of a chip, component, or subsystem impacts its reliability. Because the DLT is influencing the average lifetime, there is a Poisson distribution of the probability of these chips or components to fail over their lifetimes, with the median lifetime around the DLT. This directly impacts the reliability of the chips and the systems on which they are based. Therefore, crucial components in redundant systems should be designed and built with components having high DLTs. This is even more important for parts that are Single Points of Failure.

As we can see in this discussion, it does not help and does not make sense to create all components or subsystems within a system failsafe. It is virtually impossible to do so, and the cost would be prohibitive to come even close. The only consequence we can draw is to accept component and module outages or failures and deal with them appropriately. Knowing the probability of some components, chips, or subsystems to fail, and their impact on the system, allows us the choice to leave them as single points of failure or make them redundant. A legitimate possibility is redundancy for modules or components that are crucial to the operational status of the device. This approach to system reliability—using redundancy—must be combined with OAM&P to provide true failsafe High Availability.

DEFINITION OF AVAILABILITY

Availability is the status of a system, a subsystem, or a module in which it is able to perform its intended function under the stated conditions.

An available system performs its functions as specified, responds to commands, then sends status data back. A system that fulfills High Availability requirements is characterized by the following:

- High Availability in TelCo terms is constituted by a cumulative system downtime of less than 3 minutes a year (99.9996% system uptime).
- System downtime includes downtimes caused by unrecoverable hardware defects, software errors and traps, and software maintenance.
- To achieve High Availability status, the hardware must be capable of supporting more than 99.9996% uptime.

Consequences out of the above requirements can easily be derived and are as follows:

- All subsystems and modules must be hot-swappable, independent of whether they are redundant or Single Points of Failure. A subsystem that is not hot-swappable will force a system shutdown for the replacement to be installed. A subsystem or module that is not swappable at all will force the entire device to be replaced.
- A system capable of HA must support redundancy on a subsystem and module level for all modules and subsystems that are crucial to the fully functional status of the device.

- Hot standby (Hot Spare) or Triple Modular Redundant Systems must be supported by the subsystem and module architecture, both in hardware and in software, for all modules and subsystems that are crucial to the fully functional status of the device.
- The Hot Swap process should not affect the system data integrity as to not impact system availability during a module swap.

Most likely, this directly impacts some internal components and subsystems. A preliminary assessment of the number of internal components that have an impact on the system availability and the acceptable error rates before a link must be assumed to be malfunctioning gives the following:

- To achieve a system uptime of more than 99.9996%, mostly components with bit error rates (BER) of less than 10^{-15} are required for internal links.
- All internal communication paths (buses, links, channels) must provide BERs less than or equal to 10^{-15}.
- The BERs will have to be monitored and compared to two thresholds: one for counting errors at expected rates indicating normal conditions, and one for taking a link or a component out of operation.
- It is crucial that errors in datagrams (payload and header or LCI) are detected.
- Errors that span multiple datagrams must be detected.
- Whenever a datagram is altered, the integrity of the incoming data must be verified and a new protection must be applied.

It is important to stress one more time that HA in TelCo terms is a cumulative system downtime of less than 3 minutes a year—including hardware defects, software errors and traps, and software maintenance such as patches and version updates or upgrade installations. This can only be achieved using redundancy mechanisms. Typically, there is a component redundancy and a board redundancy, plus software support on all affected modules and subsystems, as well as on the OAM&P card for the switchover.

1+1 REDUNDANCY

1+1 Redundancy is defined as an architecture in which a non-core subsystem is connected to two core subsystems such that the two core subsystems operate in a load-sharing way while both are able to perform their intended function, and are operating as a single subsystem with the appropriately reduced performance when only one of the subsystems is able to perform its intended function. The 1+1 redundancy can be implemented as a simple dual-CPU (or dual-function) Symmetric Multi Processor (SMP) system, or any other dual symmetric architecture, and therefore is relatively cheap. Both redundant subsystems must be able to signal their load and availability status to each other in order for the system to perform load-sharing and redundancy functions. These functions can be executed in software or in hardware, but it must be ensured that the protocol is robust against violations of itself. For example, one processor must take over all incoming load if the second processor is not responding to requests to take excess load. During normal operation in which both subsystems are able to perform their intended function, each of the units operates at 50% of their intended maximum sustainable load. During an error or an outage of one of the two subsystems, the other one must take over and then operates at 100% of its intended maximum sustainable load in order to maintain the rated throughput of the redundant module. This means that each of the subsystems will have to be rated at a sustainable throughput identical to the throughput of the entire redundant module. Typically, both redundant subsystems share resources such as parts of their memory, I/O, and some semaphores. Failure of any of these can lead to an outage of the entire redundant module. In an error case, the remaining subsystem will alert the OAM&P entity of this. The OAM&P entity will then take appropriate steps.

1:1 REDUNDANCY

1:1 Redundancy is defined as an architecture in which a non-core subsystem is connected to two core subsystems such that the two core subsystems operate in a configuration with an active core subsystem and a hot or cold standby core subsystem (see Figure 6.3). Both subsystems receive the same data, process the same data, and send out the same data for as long as both subsystems are able to perform their intended functions. If any error occurs, the data that both redundant systems send out will differ. The multiplexer and demultiplexer chips and the select logic cannot determine which inputs are valid, and which are not. It will forward the input of the active subsystem to the remainder of the logic of the non-core subsystem, independent of whether it is correct or not. While this appears to be flawed, it is not. The advantage lies in the simplicity and the resulting reliability of these components that are single points of failure. The remainder of the logic

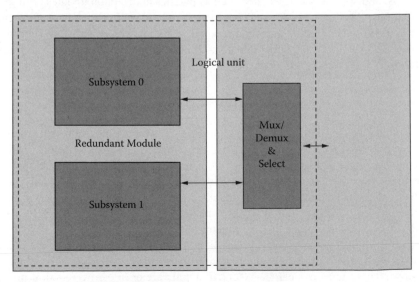

FIGURE 6.3 1:1 redundancy, hot standby.

of the non-core subsystem will evaluate the data it receives and determine its validity. In an error case, it will alert the OAM&P entity. The OAM&P entity will then take appropriate steps.

Each of the subsystems will have to be rated at a sustainable throughput identical to the throughput of the entire redundant module. This is a significantly more reliable but much more complex architecture than the 1+1 redundancy, and it provides nominal throughput even during times in which one out of the two core devices is out of service. However, it does not require any logic or software for task scheduling, load sharing, or detection of the outage in the redundant subsystem, and switchover is faster. Additionally, since there are no shared resources between the two subsystems, an outage of these shared resources cannot affect the subsystem or the redundant module as a whole. In some cases, both redundant core subsystems must work in lockstep (cycle or even microcycle synchronous).

2 OUT OF 3 (OR N OUT OF N+1) REDUNDANCY

N:(N+1) Redundancy is defined as an architecture in which a non-core subsystem is connected to N+1 core subsystems such that in a non-failure case N+1 subsystems work in parallel in a load-sharing way, and that in a failure case N subsystems work in parallel (see Figure 6.4). The performance and throughput is rated for N subsystems to be available. In a configuration with more than two subsystems an additional majority finder and decision-maker can be introduced to determine whether any one of the results is corrupted. It is assumed in these cases that the number of subsystems showing identical results is delivering the correct results, and the minority showing non-coherent results is assumed to be wrong or defective.

N:(N+1) Redundancy is not used very frequently in datacom or TelCo applications, with the exception of power supplies. For Continuous or High Availability,

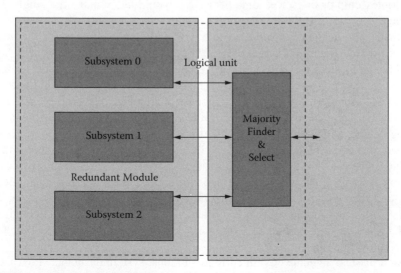

FIGURE 6.4 Triple modular redundancy, 2 out of 3 redundancy.

a 2 out of 3 redundancy in terms of power supplies is very desirable because the power supply is a single point of failure and its failure probability is not negligible. For all other applications, the additional complexity of the majority finder and the decision-maker subsystem, as the unit that determines which of the results to use and to forward, is too large, and its complexity is also contributing to a higher failure probability of this inherently crucial single point of failure. As a result, a 2 out of 3 or a 3 out of 4 Redundancy is used only in avionics and in nuclear power plants. Power supplies do not require the additional majority finder and decision-maker, and therefore this is not a complex part in a crucial path.

REDUNDANT SWITCH FABRIC CARDS

Most modern switch fabric chipsets can be used in a redundant configuration. In that configuration a fully fail-safe subsystem can be achieved. The basic config-uration of that architecture would include two switch cards with identical switch fabrics. Both of these would receive all data from all line cards, and both would process and forward the same datagrams to the same line cards, using redundant links traversing the backplane. It is in the line cards that the selection is made regarding which of the data streams to forward to the rest of the line card internal logic. The OAM&P entity declares one switch fabric card as active, and the other as the hot or cold standby. As a result, the data stream coming from the hot or cold standby switch fabric card is discarded on the line card at the multiplex and demultiplex chip using the select logic. That way, neither the switch fabric card nor the backplane or midplane will have to contain the multiplex and demultiplex chip as well as the select logic on a per-channel basis. It would be highly unde-sirable for the switch fabric card, the backplane, or the midplane to contain the multiplex and demultiplex chip or the select logic. The backplane or midplane is entirely passive, and the switch fabric card must not contain the multiplex and demultiplex chip or the select logic because it would again make it a single point of failure.

Mostly, the designer would choose a 1:1 redundancy scheme for the switch fabric card (see Figure 6.5). However, in this basic configuration the switchover would not be lossless. All cells in the active switch fabric will have to be discarded once it becomes inactive or on standby. The cells in the standby switch fabric will have to be discarded, too, since it cannot be guaranteed that both planes of the switch fabric operate in a cycle or microcycle synchronous mode. Therefore, arrival of fragmented and otherwise invalid cells is very likely. These errors will be detected and corrected in higher layers of the software on the sending and the receiving side, so this does not impose any restriction for datacom devices. In Telecom systems, however, the situation is different. Here TDM traffic is predom-inant, and there is no higher-layer function verifying Digital Service level zero (DS0). This can become especially critical if system signaling is affected, i.e., E.164 addresses or any kind of Common Channel Signaling System number 7 (CCS#7) traffic.

For a lossless switchover with current solutions, system designers either had to use non-buffered, non-queued crosspoint switches instead of buffered and queued

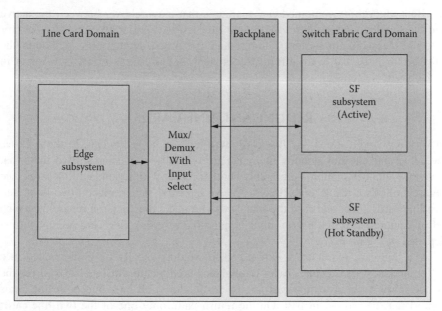

FIGURE 6.5 Redundant switch fabric cards.

switch fabrics, or they had to accept cell loss. Since non-buffered, non-queued crosspoint switches provide lower link utilization rates than buffered and queued switch fabrics, this was a less efficient solution. On the other side, a switchover involving a cell loss constitutes a significant challenge to the desired system uptime of 99.9996% or more, since not only one or a very few cells will get lost, but very likely the entire content of all buffers and queues will either get lost, will be sent by both planes and therefore destroy cell sequences in the egress port side network processors, traffic managers or SAR units, or otherwise will disable correct packet reassembly if the network processors, traffic managers or SAR units are not involved in the lossless switchover algorithm in buffered and queued switch fabrics. Either solution is a tradeoff.

However, it is possible to design a method that allows for a lossless switchover under each and every circumstance. To achieve this, the switch fabric chipset must provide lossless switchover capabilities in conjunction with the network processor or the traffic manager. In this case, the network processors or the traffic managers on the ingress and the egress sides must maintain a history. In a sliding window of a predefined size, the network processors or the traffic managers must keep copies of all sent cells and tag them as not successfully sent for as long as it did not receive explicit notification from the receiving side. Once they are tagged as sent, they can be overwritten or deleted; mostly algorithms such as this can be implemented in a ring buffer. The egress port side network processor or the traffic manager must do the same for received cells. Since, during switchover, it is impossible to prevent cells from being sent twice, it must monitor the incoming cell stream and discard cells that arrive twice. This can be done in a sliding window as well, in order to reduce the necessary processing power.

Both algorithms combined can guarantee that a switchover between the active and the standby switch fabric occurs in a lossless way, under all circumstances: for OAM&P reasons, for system failures, and for timed switchovers. Although the software and the hardware effort are not non-negligible, the effort is worth it in TelCo grade systems.

REDUNDANT LINE CARDS

In very few applications, the line cards will have to be redundant. In those cases, the active and the hot standby line card are typically connected to two links that have the same destination—preferably another router with redundant line cards (see Figure 6.6). As a result, both routers receive the same datagrams on two links into two line cards each, and then the line cards can make the decision regarding which set of datagrams to discard on input. A possible resulting line card architecture is depicted in Figure 6.7:

If a line card redundancy is required, it would typically be implemented in a 1:1 redundancy architecture. A 1:1 redundancy architecture for line cards forces the switch fabric to bicast (multicast to two destinations) all data into both line cards comprising the redundant pair. On the return path, only one of the two line cards that comprise the redundant pair would be allowed to send data through the switch fabric to the receiving egress side line card or cards. This poses quite a few challenges for the software on the line cards that constitute the redundant pair, and it also forces the router to which the redundant pair of line cards is connected to make a choice on which of the data to forward. Not only MAC and IP addresses, but also routing table information is affected by such architecture. While the MAC and IP address issue is rather simple to solve, the routing table entry must clearly specify a bicast, and on top of that, the switch fabric must allow for simultaneous bicasting.

If the transceivers for the active and the hot standby lane are aligned, a switchover should not destroy more than one bit in transit. Theoretically, a switchover can occur without even a bit loss. However, during a switchover, if one bit is corrupted, then

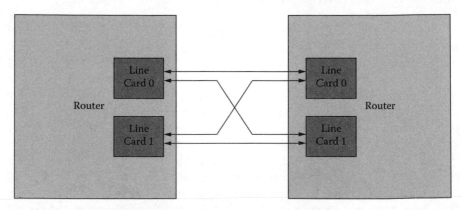

FIGURE 6.6 Redundant line cards and links between HA routers.

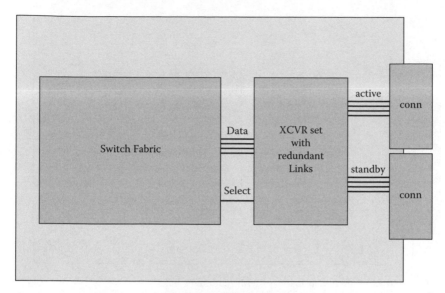

FIGURE 6.7 Switch fabric interface for redundant line cards and links.

the worst case is that it corrupts the header of a cell, therefore routing it to the wrong destination and causing a maximum of four packets to be destroyed. This is the maximum acceptable service disruption. This is not entirely true for systems that deploy High Speed Serial Links (HSSL). It is highly unlikely that the transceivers are sufficiently well synchronized between two independent links. At least the current symbol—typically an 8B/10B encoded datagram—will be lost to a switchover. If the transceivers are not both active and both sending and receiving datagrams, then it takes even more time to reacquire synchronization for the HSSL that was in sleep mode or shut down. In that case, a significant number of symbols, and therefore datagrams, will be irretrievably lost.

If a pool of line cards should share a single hot spare redundant line card in an N:(N+1) redundancy case, this can easily be accomplished by providing this extra line card and having the ingress side network processor redirect all traffic from the failed card into the standby (hot or cold standby) card (see Figure 6.8). However, a switchover will not be seamless nor will it be lossless. Neither a hot nor a cold standby line card can be current on all statuses of all other line cards, and the traffic manager's DRAM, as well as the Reassembly buffer content, are replicated amongst the line cards. As a result, all datagrams that were in transit during the switchover command will be lost; more importantly, some subsequent packets or cells will be incomplete and remain incomplete.

A possible easier way to implement line card redundancy is to make only the uplink line cards redundant, and then on the uplink line card, have the egress side network processor bicast everything towards two MACs or framers and have those links be crossed over into two routers or two line cards in one receiving router. There is no need to make the downlinks redundant.

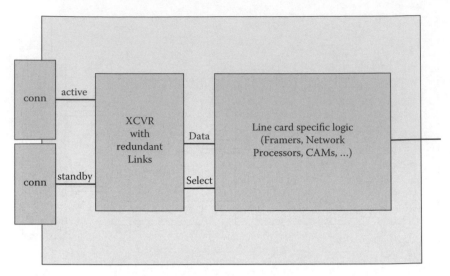

FIGURE 6.8 Non-redundant line card with redundant links.

REDUNDANT LINKS

In some cases, link redundancy is required in addition to the redundancy of core subsystems (see Figure 6.9). Especially across the backplane, link redundancy has its advantages because it alleviates the need for immediate removal of the node if the backplane has been damaged and an exchange of the node is not possible in a timely fashion. It must be said one more time that the backplane or midplane is passive. There must not be any active components on the backplane—not even electrolytic capacitors. Ceramic capacitors and resistors for line termination are the only allowed passive components on the backplane or midplane. All active components must be on the cards that plug into the backplane or midplane or other removable subsystems of the router.

Redundant links are fairly simple for parallel buses and any interconnect with a low symbol rate. In those cases, alignment and skew are under tight control, and it can be made sure that there is no more than a half-symbol time skew between the pairs of a redundant link. Alignment can be taken as a given in parallel buses with low symbol rates. As a result, a switchover becomes easy even if the links are not clocked by the same oscillator. However, this is not the case for modern routers anymore. Most modern routers deal with data rates that do not allow for traditional buses or low rate links to be deployed. Once High Speed Serial Links are used, the situation is not as simple anymore. Deskew and alignment must be guaranteed in such a way that all bits belonging to a logical link are aligned and free of skew. Additionally, the skew between any individual active line card and its hot standby counterpart should be close to zero, but at maximum one bit time. As a result, the links must be clocked from the same redundant oscillator, and their Clock and Data Recovery (CDR) units must be identical. As a result, only transceivers from the same manufacturer can be used for redundant pairs of links since every manufacturer has its own CDR design.

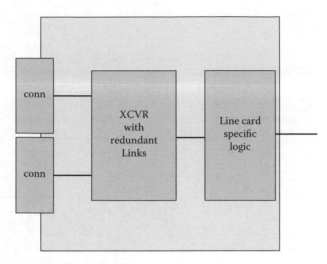

FIGURE 6.9 Line or port card with redundant links.

Independent of the specifics of the implementation of the transceiver and its CDR, it must be able to select an input channel for forwarding based on information it receives on a select signal. This requires the redundant transceivers to have the basic architecture shown in Figure 6.10.

It is important to understand that this is required for all lanes within a channel. If this is necessary in an environment with HSSLs, the logic is within the transceiver because the coding on the line requires the analog or mixed signal portion to respond to channel input, independent of whether it is considered active or hot standby.

Another issue that is different for HSSLs from traditional transceivers is the Bit Error Rate (BER). While HSSLs are probably much more robust than Low Voltage Differential Signaling (LVDS) or Low Voltage TTL (LVTTL)—the mere fact that these operate at significantly higher symbol rates and that they use 8B/10B coding means they can detect code violations and errors easier than older technologies—bit errors must be expected. HSSLs will report errors from time to time, unlike LVDS transceivers. HSSLs are monitored and must have bit error counters to determine the BER over the link. The OAM&P entity typically sets two thresholds. One threshold is used to detect decreasing line quality; the other is used to take the line out of service. So if an error occurs, the system requires the link to resynchronize and restart, but also to notify the OAM&P entity via whatever measure to report BER1. If it is below both thresholds, then the OAM&P entity stores it in a log file and continues to operate normally. If the BER1 threshold is exceeded, then the OAM&P entity will notify the NMC of a possible line down situation in the near future (known as predictive failure analysis, or PFA). If errors continue to occur then the BER2 threshold will be exceeded rather quickly. The system will take the line out of operation and will notify the operator about this issue. On a chip level, there must be the bit error counter and a possibility to generate an interrupt upon exceeding BER1 and BER2. (See Figure 6.11.)

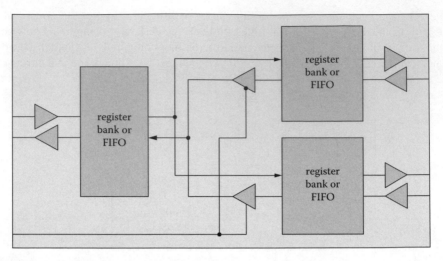

FIGURE 6.10 Redundant transceiver, in a single lane or bit parallel.

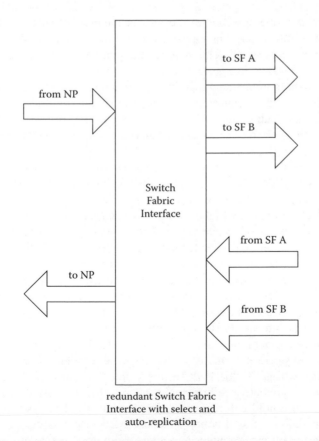

FIGURE 6.11 Redundant switch fabric interface.

The entire situation becomes a little bit more difficult in routers that must make use of parallel High Speed Serial Links. Not only are deskew and alignment between the individual lanes of a channel required to be monitored and kept under control; more importantly, the redundant parallel HSSL must be aligned and synchronized during all times as well. The availability of the redundant parallel HSSL is key to the switchover and potential loss of data.

REDUNDANT POWER SUPPLIES

For Continuous or High Availability, redundancy in terms of power supplies is very desirable because the power supply is a single point of failure, and its failure probability is not negligible. 2 out of 3 Redundancy is used very frequently in power supplies for High Availability datacom or TelCo applications.

Without any power, all other redundancies are useless; if there is no power, then none of the other measures are of any use. Therefore, a 2 out of 3 strategy for power supplies is advisable. This means that two out of the three power supplies can power the device. Action must be taken soon, because the remaining two power supplies are required to run the device. In those cases, neither of the remaining two can fail, and therefore, the time in which only those two are available must be kept as short as possible. Typically, a technician is dispatched to immediately replace the impaired power supply and restore the 2 out of 3 redundancy. Most carriers require this to be 15 minutes or less, and therefore the supplier of the Telecom grade equipment must have a service organization that can make this happen.

In case this situation persists, some of the more modern suppliers of the Telecom grade devices have implemented software in their equipment that makes sure even one power supply can run the device, albeit with throughput restrictions. So if after a preset time no service has been carried out, some of the non-essential functions go into power-save mode. Traffic management is disabled, and all DRAM associated with it is powered down. A hot standby switch fabric card will be sent into cold standby, making switchover more time consuming in the case of a failure. The hot standby OAM&P card goes into power-save mode too, but will wake immediately if something goes wrong with the active card. It just will not work anymore in lock step with the active one, if it did before. This allows the device to save enough power so that, in the case of a failure of the second power supply, its availability will not be impacted. There definitely is a throughput and a performance impact, and all traffic management is canceled, but the device is not down. It is available; it can deal with high priority traffic; it can perform everything that is required except for low latency switch fabric switchover, traffic management, and queuing large amounts of traffic. It will also be impaired in terms of OAM&P redundancy in a way that no lock step operation is possible anymore—but the second OAM&P card is still available for service in case the active one fails.

The goal is that the device can then take further measures to reduce power consumption of the remaining cards such that power-save mode can be entered, and then even one power supply can power the device in a power-save mode with a reduced device throughput. Although it is undesirable to have a throughput-reduced situation, it is still better than having a device-down situation. This must be entered only if the service call to the NMC has not been answered in a timely fashion. Typically, the setup of the OAM&P entity would be such that if a service call to replace a defective power supply is not answered within 15 minutes, the device prepares for a system-wide power-save mode so that the failure of another power supply will reduce throughput but will not bring it down.

However, if power supplies fail, not all routers support a throughput reduction along with a reduction in power consumption. As a result, it is always preferable to have the failed subsystem—especially the power supply—serviced and the 2 out of 3 redundancy reestablished.

SOFTWARE ROBUSTNESS

While we have discussed the impact of IPv6 and similar protocols to unify multiprotocol networks on the hardware of modern routers, we have not yet touched the issue of software. SLAs and QoS guarantees have an equally important impact on software and hardware. Since current systems contain a significant amount of software, making only the hardware "bulletproof" is not good enough. In order to achieve a system uptime of more than 99.999%, a few measures have to be taken to improve and maintain software quality. Not only does the designer have to make sure that software does not crash, he also has to ensure that even grossly wrong input parameters do not throw off the system. The largest contributor to system downtime is software. Therefore, the effort must be mostly on reducing software bugs that might cause the system to be unavailable. It is of no use if the hardware by itself is crash-proof but the software running on it frequently crashes or must reboot due to software faults. That means that the software itself must be capable of surviving some hardware errors, but more importantly, it must have a flawless stack and heap administration. Garbage collection must be flawless. In case something goes wrong anyway—like spurious hardware faults or nested Interrupt Service Routines (ISRs) that are terminated because of run-time super-vision to avoid stack overflow and a subsequent software crash—a planned shut-down and restart or recovery has to be implemented. However, rolling recoveries must be avoided. Additionally, all usual contributors to software crashes must be avoided. Garbage collection, heap management, release of unused stack or heap space, or any other memory allocation and deallocation schemes must be extremely robust, and it is more than advisable to carry out periodic memory scrubbing. As a result, it is advisable that these tasks are run periodically. A preferred method to ensure this is to run these tasks in the background, triggered by an external timer and not a piece of software running on the CPU.

Reducing system downtime due to software maintenance or updates is crucial to achieve High Availability status. It implies that a software update or upgrade can be implemented in the background without service disruption. Therefore, it is pre-ferred that a software update or upgrade can be implemented as a background task without impacting system availability. In this case, the total system downtime of 3 minutes per year can be allocated to actual hardware failures, unavoidable software failures, or any other situation in which the device itself is unavailable for service. This can be done in a variety of ways.

First of all, the software must support the hardware in its functions. More importantly, it must do so without sacrificing throughput, delay variation, latency, and robustness. As a consequence, it is easy to conclude that the operating system on each of the router modules will have to be a multitasking operating system with a watchdog timer; it must be able to assign different priorities to different kinds of applications. The innermost layer of software clearly must be the scheduler task, running at the highest level of priority.

Second, it must also be able to support and accommodate prioritized Interrupt Requests (IRQs) by invoking an appropriate Interrupt Service Routine (ISR) as soon as an Interrupt Request is issued by peripheral components and forwarded to the CPU. The watchdog timer typically is set such that it will trigger a reset on the CPU if it is not reset within its expiration period. The timer will usually be set such that it would expire after 512 ms, which typically equals 512 "timer ticks." The software should reset the timer every 500 ms, which it will do if it is not stuck in an infinite loop. If it is, then the main process that resets the timer will not be able to do so, and the software will be restarted by an external reset event through the timer.

In all telecom and datacom systems, software traps and hardware IRQs should be treated in the same technical way, through Interrupt Service Routines (ISRs). However, nested ISRs can become a significant problem and must be avoided. While it is absolutely appropriate to deal with a software fault within an ISR, it must be ensured that they cannot be invoked out of the invoking ISR. The same is true for hardware-triggered Interrupt Service Routines: while an ISR to an IRQ is being processed, the Interrupt Controller must ignore further IRQs from the same source, and clear the Interrupt Status Register immediately. If that does not occur, nested ISRs from IRQs and software faults will happen. It will not take long for them to exhaust all available memory. To avoid nesting ISRs, software traps and hardware IRQs must consequently be handled with the utmost attention to detail. One way to ensure this is to have the highest-priority ISR for a hardware-based RESET signal or IRQ input reset all conditions from external requestors. The watchdog timer can do this within the ISR that deals with the time-triggered reset.

This type of operating system and scheduling process will run on each CPU or microcontroller on each of the router modules. Typically, the entities impacted by this are as follows:

- Line card
  - Network processor
  - Traffic manager
  - CAM control
  - Local control CPU

- Switch fabric card
  - Local control CPU

- OAM&P card
  - Main processor
  - Local control CPU

Like any other multitasking, multithreaded operating system that has to respond to external triggers, this carries the potential problem of a rolling recovery—the CPU reboots, and gets into the same infinite loop again, cannot reset the timer, and is reset once again. Situations like this are undesirable, but can happen since it is impossible to write software that avoids it under all circumstances. It can happen that external parameters and conditions are unforeseen, may be out of range, and therefore the same erroneous environment will reemerge every single time. However, this is a very low probability. Rolling recoveries on any component in the data path, but also on the OAM&P cards, must be avoided at all costs. Therefore, not only the system test must be very thorough, but more importantly the system must be set up such that each and every individual module is as simple as possible to allow for testing a significant portion of all system statuses and status transitions.

OAM&P CONTROL OVER REDUNDANT SUBSYSTEMS

As we have seen in the OAM&P chapter, it does not make sense for the local control CPU to determine the administrative status of a managed entity. In contrast, the system-central OAM&P entity should be in control of all system internal subsystems. The only other entity involved in the determination of any operational status that can override the local OAM&P entity's decision is the NMC. As a result, the ultimate decision-maker in the system is the local OAM&P entity without NMC override. The OAM&P entity ultimately must be in control over all relevant subsystems to ensure that the system is able to perform its intended function; in case it is not, the OAM&P entity notifies the NMC. In order to do so, the OAM&P entity must be able to communicate with all other subsystems. The OAM&P entity can use data path functions, or control path functions within the data path. However, if the data path is down or the control path is down within the data path, then OAM&P becomes impossible. As a result, OAM&P communication within the router should be carried out through dedicated communications channels. It has been proven very cost-effective and more than sufficiently reliant to use Fast Ethernet as a means of communication between the OAM&P entity and the line cards, the switch fabric cards, and between potentially redundant OAM&P entities. Communications with fan trays and temperature sensors within the chassis are typically implemented with I^2C.

As we can see, High Availability is only possible with OAM&P involvement. Since there will always be module failures, only OAM&P can ensure the device as such can continue to function; in case that is not possible, only an OAM&P entity can ensure that the NMC is alerted and ready to take further action.

TIMED SWITCHOVER

In many applications, it makes sense to perform a timed switchover between the active and the hot standby component or subsystem. Typically every 12 or 24 hours, both the active and the active standby devices are switched over. That way, it can be ensured that both the active and the hot standby systems are routinely used and prove their ability to function correctly. Without the periodic timed switchover, there is no proof that the subsystem or the module is able to perform its intended function. As a result, it is useful to periodically swap the operational statuses of the active and the standby subsystems, so that the subsystem that was previously active becomes the standby subsystem, and vice versa. This is useful to make sure that the designated backup subsystem, in case of failure, indeed functions as intended. The status swap is typically performed during a period of time in which the traffic is expectedly low. Should the subsystem—previously the standby subsystem and now the active subsystem—fail or function incorrectly, all subsystems in its data path will start to receive invalid datagrams. As a result, they will alert the OAM&P entity. The OAM&P entity will in return determine if the switchover was a timed switchover and therefore can be reversed without any negative impact, or compare current error rates versus previously recorded error rates of the subsystem that is now the standby subsystem. The OAM&P entity will direct all affected subsystems to do so if the situation promises to improve with another status reversal. If the situation improves, then the OAM&P entity will declare the current standby subsystem as defective, so that a subsequent status transition to active becomes impossible, and notify the NMC of this situation. The advantage of the timed switchover is that a faulty standby system can be more easily detected, and corrective action can be taken before peak traffic occurs.

SWITCHOVER ON DEMAND

Switchover on Demand and timed switchover are not mutually exclusive—the opposite is true. For most applications, Switchover on Demand and timed switchover are combined so that there is a timed switchover that is routinely invoked every 12 or 24 hours, and a Switchover on Demand that is triggered by unplanned failures and outages. Switchover on Demand is crucial to High Availability. If there is no Switchover on Demand, then there is no reaction or response to unexpected outages and failures. As a result, the system cannot be kept running and available—able to perform its intended function under the stated conditions. Switchover on Demand therefore is crucial to the availability of the system. The importance of Switchover on Demand cannot be underestimated. While it requires functioning communication between affected subsystems and the OAM&P entity, it enables the OAM&P entity to set the administrative status of any of the subsystems to the appropriate status optimized towards system availability and not focused on the subsystem view. Switchover on Demand truly determines system status information and then determines the appropriate action—something local control CPUs cannot do, and something that the NMC will not be able to do either. The NMC response would be too late and would not take into account the specifics of the local system.

REDUCING HUMAN ERROR PROBABILITY

While this sounds somewhat obvious, it is not. Unclear and ambiguous signals, status displays, and even mechanical arrangement can make the identification of a particular subsystem or module difficult or impossible. As a result, all modules should be clearly marked. Counting of modules should start at the left side and the numbers should increase going to the right. Numbers should be consecutive if there is any potential of a misunderstanding. All subsystems that are in working order and are declared ACT by the OAM&P entity should have a green LED lit. A defective module should display a red LED lit. If hot standby modules are present and the intention is to signal that they are not ACT but STB, then they should display a blinking green LED. Modules in MBL could display an orange LED lit. It should be verified that using the levers to pull out one module does not unintentionally pull out another module, even by a technician with very large hands. A reset button on a module should be placed far away from levers and any other component that requires mechanical intervention when pulling out or inserting a module. Preferably, if there must be a reset button on a module, it should be located behind a flap or it should be required to press it for more than three seconds to activate it. When activated, it should clearly display that a reboot is in progress. At the same time, the software should disable a subsequent reboot to avoid rolling recoveries induced by mechanical operation of the reset button.

CONCLUSION

High Availability might be required for a significant number of advanced routers. This can be due to deployment in mission-critical networks, within the Internet2, within corporate backbones and other core networks, or it could be required for routing PSTN Emergency phone and cell phone calls. In any case, High Availability requires an integrated solution, consisting of hardware and software within the router, OAM&P entities, and a Network Management Center (NMC) that has a network-wide view of node statuses and therefore can make more useful decisions than a local device alone. Switchover on Demand and timed switchover help the OAM&P entity to maintain the system's health and ensure that the node is in working order. The hardware and software within a modern router that can help increase the system availability is based on hot-swappable modular subsystems that eliminate Single Points of Failure as much as possible. Robust software relies on better code review techniques, on watchdog timers for all mission-critical modules, and on garbage collection techniques that periodically will have to scrub all relevant portions of memory.

The last portion of High Availability is minimizing the influence of human error. This can be partially achieved by clarity and by eliminating ambiguity. It is of absolutely no use if a router has the perfect redundancy scheme and a hot standby switch fabric card losslessly takes over from an active card that failed—and the service technician is dispatched and removes the now-active switch fabric card due to a *red* (instead of a green) LED indicating operational status.

7 The Chassis

OVERVIEW

Current routers come in single-board designs ("pizza boxes"), midplane designs, and backplane designs. Very likely, none of that will change with the new generation of routers. However, the advantages and disadvantages of the three designs today are very pronounced, and they will be even more pronounced with newer router designs.

This has largely to do with the fact that the deployment of the next generation of routers will be much broader than it is today. Today, routers have taken over the corporate data network. However, they are not predominant in the Central Office yet. Therefore, neither density nor failsafe design nor serviceability while in operation have been high priority design considerations. This will change in the very near future, and for that reason routers in COs will take over a lot of the mechanical designs from TelCo gear, such as Class 4 and Class 5 Central Office Plain Old Telephony System (POTS) switches. For routers that are not deployed in these categories, the mechanical design will focus more on compactness and price, therefore resulting in a different category of products.

From the Chapter, "Internet Topology Change," we can directly take the requirements for future IPv6 and other advanced routers. They will have to

- Losslessly switch and route datagrams
- Be able to perform Segmentation And Reassembly (SAR)
- Perform policing at line speed for thousands of virtual connections simultaneously and thereby enforce and support SLAs
- Perform Traffic Management by queuing and buffering traffic according to SLAs, and drop excess traffic
- Support Traffic Engineering by means of collecting Statistic Traffic Data
- Support a mixture of hierarchical and mesh interconnect infrastructure in terms of data traffic
- Support an overlay network for metadata and signaling data
- Enable traffic rerouting at the edge and provide redundant fail-safe systems towards the core
- Communicate securely within the components of the router
- Communicate securely between the OAM&P card and a Billing Center
- Communicate securely between the OAM&P card and a PKI Center for authentication
- Communicate securely between the OAM&P card and a Network Management Center (NMC)
- Communicate with the PSTN infrastructure

Some of these requirements have an indirect impact on the chassis design. For example, it is crucial to provide enough area on Printed Circuit Boards (PCBs) to accommodate placing Integrated Circuits (ICs) for all functions required. While the integration density of ICs tends to increase, the functions required at ever-increasing line rates still demand some PCB area to be implemented. Apparently, not all routers will have to perform all these functions. It is unrealistic to assume that an unmanaged single-board router will have to or is expected to perform traffic management and billing functions. One can assume, however, that there is a need for a significant amount of PCB area on line cards and switch fabric cards in routers that must be redundant and failsafe and perform packet forwarding at very high line rates with a utilization close to 100%. Since all routers consume a considerable amount of power, they must dissipate it. As a result, they must use the Printed Circuit Board (PCB) as a thermally conductive device. The glass in the PCB dissipates the heat generated in the ICs into the PCB and through them into the chassis.

SINGLE-BOARD VERSUS MODULAR DESIGNS

In general, routers can be classified into one of two design categories: single-board or modular. Single-board designs typically are called "pizza boxes," whereas the modular designs consist of midplane or backplane architectures. Their generalized deployment scenarios differ significantly, and their prices differ. Single-board designs typically trade the ultimate throughput, reliability, and serviceability for cost. Another distinguishing factor would be convection cooling plus forced cooling (in modular systems) versus forced cooling only (in single-board systems). For fault tolerance, serviceability, and access to cabling and ducting, modular designs beat single-board routers and switches—something that is of crucial importance to carriers, ISPs, and large data centers. In terms of cost or even switched gigabit per second per dollar, single-board designs cannot be beat. Each one of the designs is a highly evolved technology, and with IPv6 support even in the Small office/Home office (SoHo) realm, a unified network infrastructure can become reality. Modern microelectronics has packed a great number of transistors and functions onto a single chip, so it was only a matter of time before the first single-board routers were created and sold. Since all functions were and are integrated onto a single board, manufacturing has become easier and cheaper than for chassis-based modular systems at equivalent throughput rates. Mass-deployed routers with throughput rates that are not leading-edge, but are based on mass-produced semiconductor technology, are mostly single-board routers. For better prices, they offer the same feature and function set at equivalent throughput rates as modular routers. While the focus of single-board designs clearly is on providing advanced functions at a low cost, the modular designs' highest priority is high availability at high throughput. Both modular designs share a variety of design criteria, which will be the subject matter of the next paragraph.

MODULAR DESIGNS

Chassis-based architectures are predominant in applications where it is not yet possible to integrate all functions into one board, or where system uptime requirements make redundant and modular systems a necessity. Chassis-based routers fall into two categories: backplane and midplane architectures. These again subdivide into systems that have a shared bus to interconnect the line cards, or a switch-based interconnection. Routers with modular architectures share a variety of design elements. They both are deployed in environments in which high availability and high throughput are of crucial importance. As a result, they must be designed such that they provide at least some sort of redundancy to achieve the high availability requirement, they must be serviceable, and they must allow for removal of components or modules without significant impact on the system state itself. As a result, they are typically designed to occupy a rack or parts of it, and to allow for easy cabling, cooling, and other access. While single-board routers come in a variety of sizes, rack-mounted modular routers come in 19″ and 23″ width varieties. Their heights vary from 6U to full 8′ (foot) racks. Typically, they are also deeper than single-board designs, so line cards, switch fabric cards, and the OAM&P cards can be significantly deeper, which provides more PCB area on the PCB and makes heat dissipation and layout a lot easier (see Figure 7.1). In other words, the major

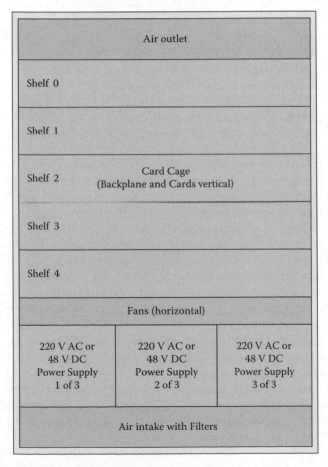

FIGURE 7.1 Typical modular design.

advantage of the racks is to provide more PCB area, allow for better cooling, and provide accessibility in case of service to any of the modules.

In all modular designs the backplane and the midplane are currently the limiting elements of throughput growth. This restriction results from the fact that electrical connections are more throughput-limited than optical connections, and that the insertion and extraction forces have to be limited to manageable values. In modern High Speed Serial Links, one link can carry symbol rates from 3.125 Gbit/s to around 11 Gbit/s. However, it must be seen as four pins on a connector per direction: +D, -D, two GND to shield. As a result, a bidirectional link with 3.125 Gbit/s symbol rate requires eight pins. The 3.125 Gbit/s symbol rate most often translate into 2.5 Gbit/s net data transfer rate because of the 8B/10B coding used. A modern router with a queue manager and a traffic manager on the line cards will always provide some internal speedup between the line cards and the switch fabric card, and therefore will require a higher data rate provided internally than is available for the external links. Typically, a factor of two is used to avoid Head-of-Line blocking and to accommodate for internal signaling overhead; therefore, 16 pins for a total of 5 Gbit/s internal capacity between the line card and the switch fabric card will be required for a net 2.5 Gbit/s external link. With a maximum of 1000 pins per connector into the switch fabric card this will provide enough capacity for 62.5 ports—not counting any other signals that might be required. Considering power and GND and other signals required, a reasonable number is 32 external ports of 2.5 Gbit/s net data transfer rate. In other words, with current HSSLs we can build a router with 32 OC-48 ports, or with eight OC-192 ports. If we want to achieve a higher number of ports or higher data rates per port, we will have to increase the number of pins on the connector into the switch fabric card or increase the symbol rate on the HSSL. Implementing 16 ports of OC-192 with current HSSLs requires segmenting the connector into the switch fabric card into the signal portion with more than 1000 signal pins, and using a second connector for power and GND. However, this will significantly increase the insertion force of the card into the card cage and stress the back- or midplane.

Another solution is to use HSSLs with a higher symbol rate. These have become available very recently, and provide 11 Gbit/s of symbol rate using 64b/66b coding. As a result, a bidirectional link carrying 20 Gbit/s for an external line rate of 10 Gbit/s net data rate requires 16 pins under the same circumstances as in the above example. This would provide enough internal throughput to support roughly 32 OC-192 (or 10 GbE) line cards, or 8 OC-768 line cards. A third solution is to limit the required speedup between the line cards and the switch fabric card by placing the traffic manager on the line card, but the queue manager on the switch fabric card— where it logically belongs.

The data communications standard for racks is 19″, and therefore most currently available IPv4 routers are in 19″ chassis. This makes sense, since they are most likely integrated with the servers and storage devices and all other machinery required to keep a data center running. In a 19″ rack, there is space for 16 to 18 cards—most cards occupy a 1″ slot. Very space-efficient chassis leave space for 18 cards; in datacom requirements this translates into 16 line cards, one switch card, and one OAM&P card. This is acceptable in the datacom world, and in data centers where dynamic rerouting of data traffic in case of a device failure can easily be accomplished. However, if NMC-controlled rerouting around a defective device is not an option, or if the requirements for system uptime are higher, the said 19″ chassis in the 19″ rack does not offer enough space. 16 line cards are desirable since switch fabrics nearly always are designed to support 2^n line cards. Scheduling and load balancing is easiest that way. Consequently, 16 line cards are desirable. In a typical High Availability scenario, two switch fabric cards are required, and two OAM&P cards are called for. This brings the total number of required cards in the chassis to 20. This, by definition, does not fit into a 19″ rack or chassis. Therefore, the Telecom industry has favored 23″ wide racks all along. These racks leave enough space to install up to 22 cards of 1″ width each. This gives the designer of a router or an ATM switch the desired 16 line cards, two redundant—ideally Hot Standby— switch fabric cards, and two redundant—ideally Hot Standby—OAM&P cards.

Large chassis in 19″ or 23″ racks have other advantages. They typically provide significantly more space for power supplies and fans. Since it is desirable to use natural convection, the airflow will go from bottom to top. No horizontal obstructions are allowed, and in most cases Telecom devices have two fan trays with three or six fans each. There are also typically three power supplies for the device, and out of those three, two are required to deliver the current the device requires. All work is in load-sharing mode, so if one fails, it does not impact the operation of the device at all, excepting that the OAM&P card will send a notification to the NMC. This redundancy in terms of power supplies is very desirable because the power supply is a possible point of failure with device-wide impact. Without any power, all other redundancies are useless; if there is no power, then none of the other measures are of any use. Therefore, a 2 out of 3 strategy for power supplies is advisable. This means that any two out of the three power supplies can power the device at full traffic load. In case this situation persists, the device can then take further measures to reduce power consumption of the remaining cards such that power-save mode can be entered, and then even one power supply can power the device in a power-save mode with a reduced device throughput. Although it is undesirable to have a throughput-reduced situation, it is still better than having a device-down situation. This must be entered only if the service call to the NMC has not been answered in a timely fashion. Typically, the setup of the OAM&P card would be such that, if a service call to replace a defective power supply is not answered within 15 minutes, the device prepares for a system-wide power-save mode so that the failure of another power supply will reduce throughput, but will not bring it down.

An additional benefit of a large chassis is that it acts as a heat sink if there is a thermal connection between the chassis and the heat sources on the PCBs. Thermal conductivity can be used in the PCBs and the chassis to divert some of the heat into the metal structure of the chassis; this increases the surface area of the heat sink and therefore simplifies heat removal. The chassis usually is equipped with multiple temperature sensors that perform temperature measurement and send the results via an I²C bus interface to the OAM&P card. While the chassis might not directly be affected by the Network Equipment Building Standard (NEBS), the rack typically is. NEBS defines a set of requirements for racks regarding earthquake safety, electrical safety, mechanical dimensions, and other parameters in the CO environment.

As a consequence, a router that fills an entire rack must comply with NEBS. A larger chassis also makes the servicing of modules easier by allowing slides to remove and insert modules and components. The levers on the chassis will have to take into account the significant forces that are present in routers having connectors with high pin counts to the backplane. The insertion and removal forces of cards with more than a thousand pins connecting the card to the backplane will require levers to be installed so that the service technician is able to remove and insert cards. Some modular routers have local control panels and displays so that the device status and the status of its components are visible to the service technician without the use of computer. In that case, the OAM&P card receives the panel input and controls the status LEDs, mostly though I²C.

The following schematics contain a few select sample backplanes or midplanes for modular routers (see Figure 7.2). The schematic shows a backplane for a modular 10-slot router that has no redundant switch fabric cards and therefore does not fulfill High Availability criteria. It has one switch fabric card, eight line cards, and one OAM&P or management card. All line cards are connected to the switch fabric card, and all cards—including the switch fabric card—are connected to the OAM&P card. The OAM&P card has an external Ethernet or Fast Ethernet connector for communication with the NMC. It uses a very common 8-port switch fabric, which may be a shared memory switch or a crossbar-based switch.

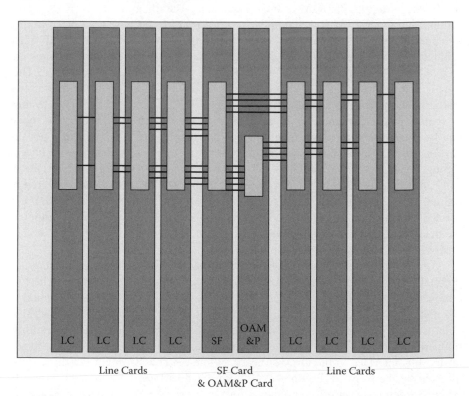

FIGURE 7.2 Sample 10-slot non-HA backplane schematic.

A slightly larger but otherwise identical version of this design is a 14-slot non-HA system (see Figure 7.3). It is used quite often if a 10-slot router does not provide enough line cards and ports. It deploys a very common 12-port switch fabric. This keeps the cost down while maintaining the modular design and thereby providing serviceability and flexibility in the choice of the line cards

Another variation of the 10-slot router is the 6-slot version (see Figure 7.4), either with 4 line cards and 4 ports, a version with quad-port cards providing 16 ports, or a concentrating version with 12 downlinks and 1 or 2 aggregating uplinks on three quad-port cards and one single- or dual-port card for the aggregated uplink. As a result, it either uses a four-port, a 12-port, or a 16-port switch fabric. Multiport cards combine several usually identical ports on one line card, and thereby reduce cost over single-port cards.

The schematic for the 18-slot High Availability backplane (see Figure 7.5) shows a backplane for a modular 18-slot router in a 19" rack that has two redundant switch fabric cards and therefore fulfills High Availability criteria. It has two switch fabric cards, fourteen line cards, and two OAM&P or management cards. All line cards are connected to the switch fabric card, and all cards—including the switch fabric cards—are connected to the OAM&P cards. Both OAM&P cards have an external Ethernet or Fast Ethernet connector for communication with the NMC. It uses a very common 16-port switch fabric. It is restricted to 18 slots by the available space in a 19" rack, but the redundant switch fabrics support 16 line cards instead of the 14 that can be installed. Sometimes, these designs support two dual-port line cards so that 16 ports are available in 14 slots.

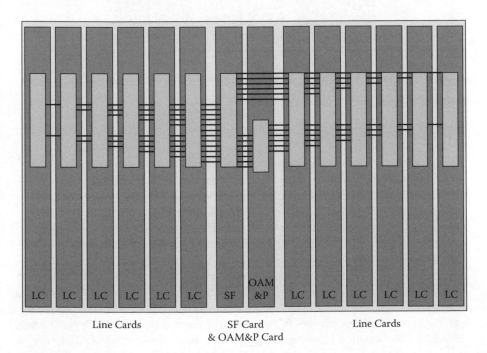

FIGURE 7.3 Sample 14-slot non-HA backplane schematic.

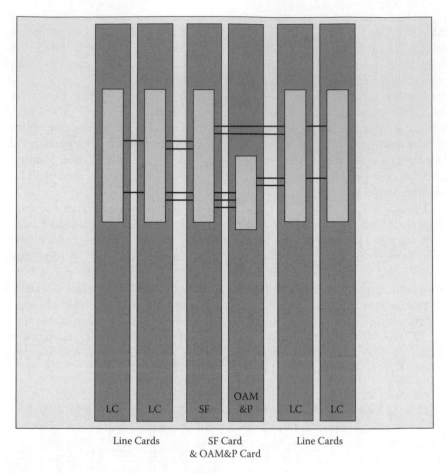

FIGURE 7.4 Sample 6-slot non-HA backplane schematic.

Power Supply Considerations

In most cases, the power supplies or power converters are down converters from 220V AC or –48V DC to 5V DC or 3.3V DC. All other conversions will have to occur on the line cards and the switch fabric cards, or any other card that needs different supply voltages. This occurs for a variety of reasons. One of them is that it is nearly impossible to provide all required voltages out of a single power supply and to foresee what requirements updates and upgrades will bring. The other requirement is to deal with losses. At 220V and 16A, the total power supply ingress power is 3520W. Assuming a 90% efficiency of the power conversion, egress power out of the power supply is 3168W. The remaining 352W unfortunately heat up the router. However, at 12V DC output, we must feed 264A into the router. This will require very wide and thick power layers on the PCBs, and will absolutely be a challenge to any connector. The situation is worse at 5V, where it must transmit 633.6A; at 3.3V it must handle 960A. In any case, all the low voltage power supply lines will require

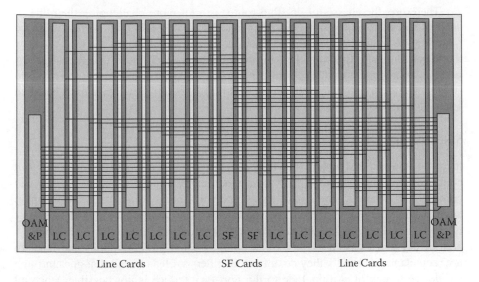

| OAM &P | LC | LC | LC | LC | LC | LC | LC | SF | SF | LC | LC | LC | LC | LC | LC | LC | OAM &P |

Line Cards   SF Cards   Line Cards

FIGURE 7.5 Sample 18-slot HA backplane schematic.

close to zero Ohm resistance. A 0.1V drop will result in 26.4W, 63.4W and 96W loss across the power supply lines and planes alone. This is undesirable for two reasons: power is cut towards the router components, and more heat is generated. More generated heat means more required cooling, which is even worse. Electric heating has a very high efficiency. Refrigeration at best achieves 30% efficiency. In other words, for every 1W generated in additional heat, at least 3W must be spent to cool it down. The real incentive of power savings are reflected here: it is not only direct savings, it is about a factor of three that can be saved on cooling the equipment. The biggest challenge to increase switching, routing, processing, and forwarding density is to get heat out. In other words, in order to increase the Mega Instructions Per Second (MIPS) and Dhrystones per cubic meter, designers must think about how to reduce the W/MIPS or W/Dhrystone or W/(Gbit/s), W/(packet/s) down to the minimum number defined by thermodynamics and entropy. The real challenge will be energy efficiency to be built into routers so that they do not melt down if switched on. Partial power down, gated clocks, variable clock frequencies and other measures to decrease power will enable denser packet processing, forwarding, and switching.

The chassis must support all of this. It must be designed such that it allows all the internal components to function properly, supply power to these components, interconnect them, and provide for cooling. The more efficient the internal layout for cooling and convection, the higher the reliability and total system availability. Not only must hot spots to be avoided on a PCB level, the PCBs must be placed such that they support convection cooling—so that all components on the PCB can function in the temperature range for which they have been designed. If gated clocks must be used to maintain operating parameters of the chip, then the router is unable to sustain throughput under all conditions. This would show up as spurious and probably unidentifiable errors or performance degradation in operation.

Therefore, the chassis of an IPv6 router is a lot more challenging than that of an IPv4 router. In addition, they will be stacked, and therefore the airflow must be quite high to maintain a low enough temperature gradient within the rack to allow the upper routers to function. This in turn might very well mean that the airflow is not smooth anymore; it will be turbulent within the chassis and within the rack. So fluid dynamics does not entirely describe the scenario, and turbulence must be taken into account.

SINGLE-BOARD DESIGNS

It can be assumed that routers for Small office/Home office (SoHo) applications will continue to be "pizza boxes," but with a much more pronounced focus on cost and compactness. These systems have no components the user can maintain or repair, and very few the user can exchange in the case of a module defect. They are very typically not High-Availability designs, and therefore not intended to be running 24 hours a day and 7 days a week without outages. Their power supplies and fans are not redundant, and mostly they do not use air filters to clean the air that goes through the device to cool it. An upgrade to the port or interface is not possible, since all interface components, processors, and the switch are soldered onto the board. The only upgrade that is possible is the upgrade to the processor(s) FLASH that contains the operating system and all application programs. No user-serviceable parts inside the device also mean that typically no component is socketed. Although this might look like a sub-standard design, it is not. It provides a lot of functionality at the lowest possible price because manufacturing is possible at a very low cost. It in fact allows SoHo users to deploy the newest technology at reasonable price points, providing the highest possible throughput with Quality of Service (QoS) guarantees for each of the traffic classes. A lot of engineering effort is put into the mechanical design of these chassis to provide a low-cost solution with a small footprint, acceptable airflow, and a design lifetime of at least three years. In order to accomplish this, the design engineers will have to use the mainboard's PCB as a thermally conductive device, designing the airflow through the device carefully to ensure that the forced cooling mechanisms are not disturbed too much by the placement of large components around heat sources such as processors, SRAM, and the power supply. The chassis itself can be made out of injection-molded plastic or stamped metal, holding only the mainboard and providing openings for the connectors of interfaces, the power supply, and the fan. Typically, single-board designs with a single large mainboard have the PCB arranged horizontally, blocking thermal convection for cooling purposes. To compensate for the lack of thermal convection cooling, forced cooling is typically used, and a reasonable design will incorporate a few temperature sensors on the mainboard and near known heat sources. The operating system will monitor these temperature sensors and shut down the router or some components if preset thresholds are exceeded.

Single-board designs offer the least space per functional block, and therefore must be designed and architected with some trade-offs in mind. In general, it can be assumed that functions that are not crucial to the operation of the device will be omitted. While every single-board router will have MACs and PHYs and a switch control—even if it is just an arbiter on a shared bus—it might omit all traffic management and most of the OAM&P functions. It will require a packet parser and a classification engine, but if space does not permit the implementation of a CAM to speed up the lookup, it might have to perform the lookup in software on the CPU, thereby restricting the number of routed packets per second. As a result, it might reach the rated throughput only for long packets. This is a direct implication of the physical size of the device.

However, with an ever-increasing integration density of Integrated Circuits and therefore a higher density of functions per unit of area, the disadvantages of a limited PCB area of PCBs becomes less and less of an issue. Current network processors and CAMs already replace a wide array of dedicated components at comparable rates of packet processing performance. The increase in line speeds and in required feature richness may counteract that density increase over time.

While single-board designs do not have the same constraints as modular designs regarding connectors between the line cards and the switch fabric card, they are constrained by the available space for components.

MIDPLANE DESIGNS

Midplane designs and backplane designs are typically used for routers that are deployed in corporate data centers or Central Offices. The midplane design is preferred when there is a common processor card design for a wide variety of line or interface cards. This makes it possible to manufacture one processor card that is common to all line or interface cards, but which can be programmed to process the different frame, cell, or datagram formats of the line or interface cards. By doing this, the designer and manufacturer of a midplane design can distribute the hardware design cost of the processor cards over multiple and different line or interface cards. It also offers the advantage that in the case of the failure of a processor card, all cabling on the line or interface card can be maintained. As with backplane designs, all PCBs in the midplane design are arranged vertically, and therefore allow for some thermal convection even if, for a brief period of time, the fans are not working.

Routers with a midplane architecture (see Figure 7.6) typically come in 19″ racks, seldom in 23″ racks. The rack contains the power supply unit, at least one fan tray, the card cage with the bolted-in midplane in the center, one or two switch fabric cards, n line or interface cards, and n processor cards. Optionally, there is one OAM&P card. Redundant OAM&P cards in midplane designs are very rare.

The midplane—which connects the port, interface, or line cards to the processor cards, and in turn connects the processor cards to the switch fabric card(s)—is bolted to the center rails of the card cage inside the chassis. As a result, the midplane is non-removable. In case it fails, the chassis must be exchanged. This results in considerable downtime unless a complete spare is available.

Midplane design routers are typically installed in temperature-controlled environments. Nevertheless, a reasonable chassis design will incorporate a few temperature sensors in the fan trays, the power supplies, and near other known heat sources. The OAM&P card will monitor these temperature sensors and first attempt to

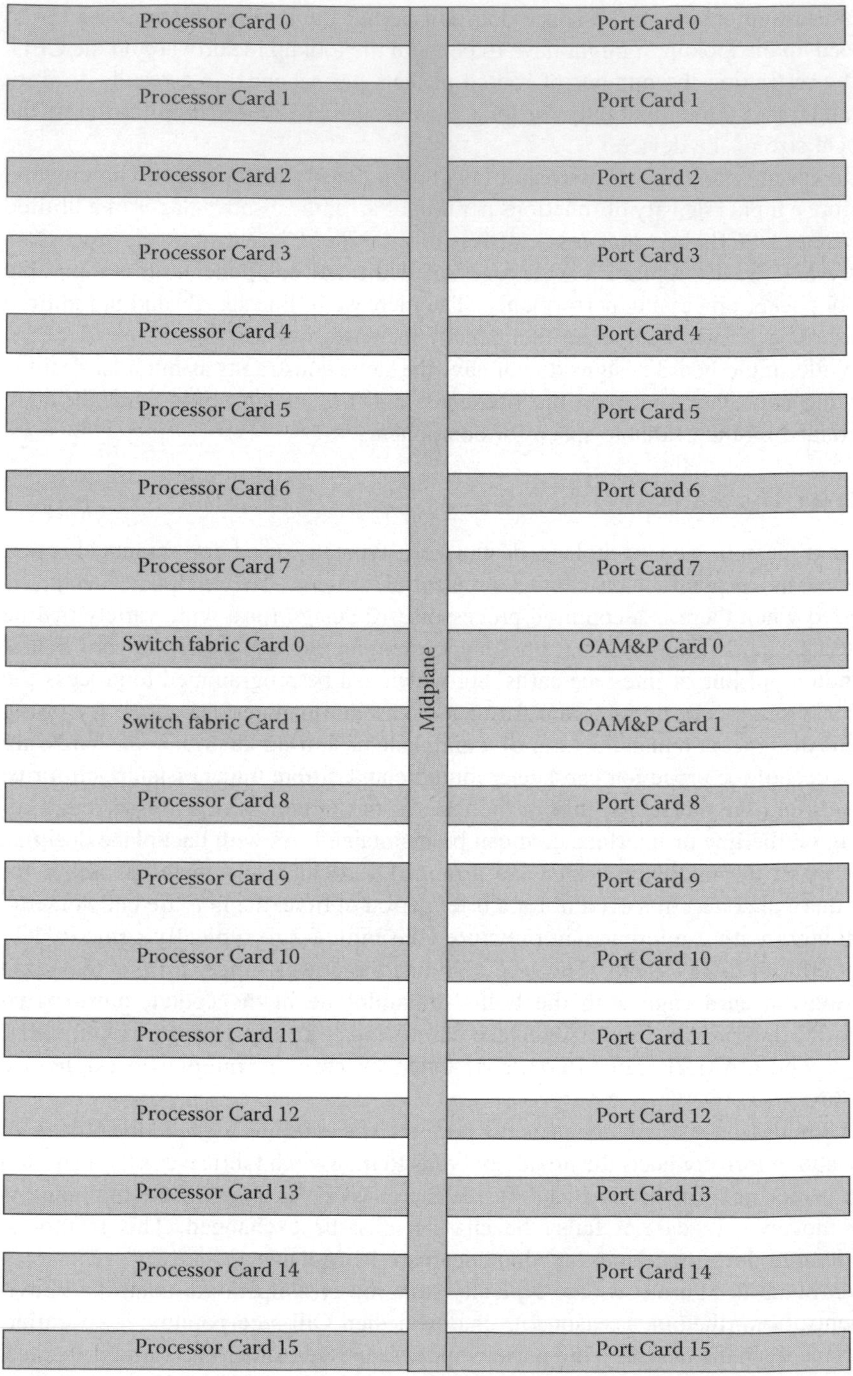

Processor Card 0		Port Card 0
Processor Card 1		Port Card 1
Processor Card 2		Port Card 2
Processor Card 3		Port Card 3
Processor Card 4		Port Card 4
Processor Card 5		Port Card 5
Processor Card 6		Port Card 6
Processor Card 7		Port Card 7
Switch fabric Card 0	Midplane	OAM&P Card 0
Switch fabric Card 1		OAM&P Card 1
Processor Card 8		Port Card 8
Processor Card 9		Port Card 9
Processor Card 10		Port Card 10
Processor Card 11		Port Card 11
Processor Card 12		Port Card 12
Processor Card 13		Port Card 13
Processor Card 14		Port Card 14
Processor Card 15		Port Card 15

FIGURE 7.6 Midplane and active cards.

increase forced (fan) airflow if preset thresholds are exceeded; if that fails, it will attempt to reduce the temperature and shut down the router or some components after having informed the NMC of the situation.

The Card Cage

The card cage contains all active components of the router and the bolted-in midplane in the center. It contains the rails in which the line or interface cards, the processor, the switch fabric cards, and the OAM&P card(s) slide into the cage. Typically, a midplane design (see Figure 7.7) holds room for one or two switch fabric cards, n line or interface cards, and n processor cards. Sometimes, space for one (rarely ever two) OAM&P card is provided. By definition, the cards in the midplane designs are significantly smaller than in backplane designs: They are only half the depth of backplane designs. This is ultimately the reason why midplane architectures for routers do not extend into the highest performance range. The available PCB area for the switch fabric is not large enough, and the switch fabric must be located on one single card. The card cage cannot reasonably be made big enough to hold full-sized single stage switch fabric cards.

In addition to this, the chassis and especially the card cage must be mechanically stable. The midplane is bolted into the chassis, and the insertion and extraction forces for cards are significant; dependent on the number of pins and the type of connector, they can exceed 1000 Newton (N). The chassis must be able to withstand these forces without any impact to its structural integrity. It will have to hold the midplane safely in place. To make things even more complex, the chassis typically is held in place only on its front panel, which in turn is screwed into the rack rails. With depths of the chassis exceeding 60 cm (24 inches) the lever arm can be significant for insertion and extraction forces of cards. It is important that the chassis does not flex or bend during insertion or extraction of cards because of the bolted-in midplane. Any flex or bending would likely cause the midplane to bend or flex, too. This might result in hairline cracks in the midplane or the connector coming loose, since connectors are pressed in and not soldered. Since the chassis and the midplane are considered non-replaceable, and both are the only single points of failure, any potential damage to the midplane must be avoided.

It must be said one more time that the midplane is passive. There must not be any active components on the midplane—not even electrolytic capacitors. Ceramic capacitors and resistors for termination of lines are the only allowed passive components on the midplane. All active components must be on the cards that plug into the midplane or other removable subsystems of the router. (See Figure 7.8.)

Because the midplane carries the physical connections of the control and data path, one might think that OAM&P signals can easily be transported using the control and data path. While, from a pure data rate standpoint, this is true, it is not true from the standpoint of reliability. Technically, OAM&P can use data path functions or control path functions within the data path. However, if the data path is down or the control path within the data path is down, then OAM&P becomes impossible. As a result, OAM&P communication within the router should be carried through dedicated communications channels. It has been proven very cost effective and more

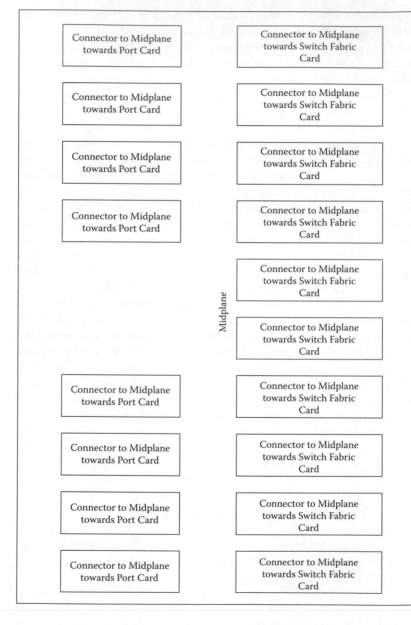

FIGURE 7.7 Basic midplane schematic.

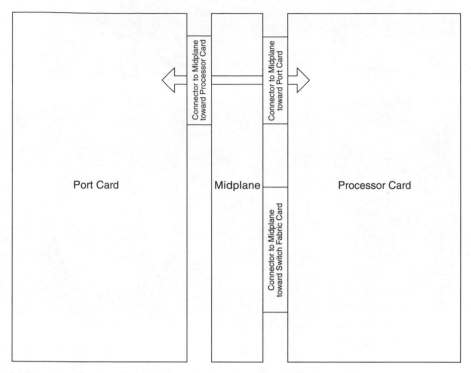

FIGURE 7.8 Interconnect between processor and port cards.

than sufficiently reliable to use Fast Ethernet as a means of communication between the OAM&P entity and the line cards, the switch fabric cards, and in between potentially redundant OAM&P entities. Communications with fan trays and temperature sensors within the chassis are typically implemented with I²C.

Power Supply Unit

Midplane designs—like backplane designs—typically feature redundant power supplies. These may be two power supplies that operate in a load-sharing mode, or three power supplies that are rated such that any two of them can power the router. For routers with power consumption below 1600W, local line voltage is the typical input: 110V AC in the US, 220V AC in the rest of the world. Some routers colocated with Central Office equipment and that need to make use of the Uninterrupted Power Supply (UPS) into the CO will require a –48V DC power converter. All power supplies must be hot-pluggable. It is very useful for the power supplies to have temperature sensors and fan speed control circuitry. Typically, the power supply would have an I²C bus interface to the OAM&P card.

Fan Trays

Typically, a midplane design router would feature two fan trays (see Figure 7.9) with six fans each; one below the card cage; the other above the card cage. The fans are

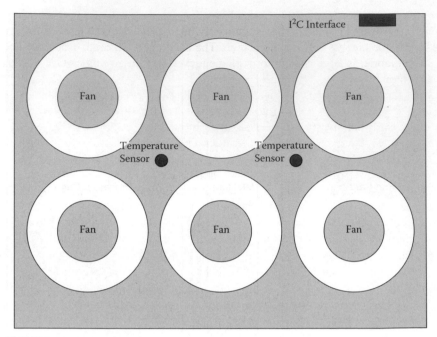

FIGURE 7.9 Fan Tray

set up such that they support natural thermal convection. A brief removal of a fan tray to exchange a nonfunctional fan therefore does not significantly impact the temperature inside the chassis. The fan trays typically support fan speed control and temperature measurement via an I²C bus interface to the OAM&P card. Fan trays and the embedded fans are hot-pluggable.

BACKPLANE DESIGNS

Backplane design type routers are typically installed in temperature-controlled environments. Since the power consumption and therefore the heat dissipation of the modules is significantly higher than in other architectures, a reasonably designed chassis will incorporate a few temperature sensors in the fan trays, the power supplies, and near other known heat sources. The OAM&P card will monitor these temperature sensors and first attempt to increase forced (fan) airflow if preset thresholds are exceeded, and if that fails to reduce the temperature, shut down the router or some components of it, after having informed the NMC.

The Card Cage

The card cage contains all active components of the router as well as the bolted-in backplane in the back of the card cage. It contains the rails in which the line or interface cards, the processor cards, the switch fabric cards, and the OAM&P card(s) slide into the cage. Typically, a backplane design holds room for one or two switch fabric cards and *n* line or interface cards. Space for one or two OAM&P cards is sometimes provided. All cards in the backplane design are significantly larger than in midplane designs. The cards are nearly as deep as the card cage and therefore provide significant PCB area. This is ultimately the reason why backplane architectures cover the highest performance range. The available PCB area, especially for the switch fabric, is large enough to implement even very large single-stage switch fabrics on one card.

In addition to this, the chassis and especially the card cage must be mechanically stable. The backplane is bolted into the chassis, and the insertion and extraction forces for cards are significant; depending on the number of pins and the type of connector, they can exceed 1000 N. Considering that backplane design allows for the largest switch fabric cards, they will have by far the highest pin count between the switch fabric card and the backplane. The chassis must be able to withstand these forces without any impact to its structural integrity. It will have to hold the backplane safely in place. To make things even more complex, the chassis typically is held in place only on its front panel, which in turn is screwed into the rack rails. With depths of the chassis exceeding 60 cm (24 inches) the lever arm can be significant for insertion and extraction forces of cards. It is important that the chassis does not flex or bend during insertion or extraction of cards because of the bolted-in backplane. Any flex or bending would likely cause the backplane to bend or flex, too. This might result in hairline cracks in the backplane or the connector coming loose since connectors are pressed in and no longer soldered. Since the chassis and the backplane are considered non-replaceable and both are the only single points of failure, any potential damage to the backplane must be avoided. In contrast to midplane designs, the connectors for the switch fabric cards and the line cards are on one side only. This allows some mechanical reinforcement of the backplane that is impossible to achieve with midplane designs (see Figure 7.10).

It must be said one more time that the backplane is passive. There must not be any active components on the backplane—not even electrolytic capacitors. Ceramic capacitors and resistors for termination of lines are the only allowed passive components on the backplane. All active components must be on the cards that plug into the backplane or other removable subsystems of the router.

If the routers are deployed in Central Offices, they will have to be rack-mounted, and they likely will have to use 23″ racks since that is the usual standard in TelCo system. Support of the cabling infrastructure and accessibility of the modules for servicing become important items. As a result, the cabling will have to be at the rear of the router. Typical CO applications require service access to the front and all cabling at the back.

Because the backplane carries the physical connections of the control and data path, one might think that OAM&P signals can easily be transported using the control and data path. While, from a pure data rate standpoint, this is true, it is not true from the standpoint of reliability. Technically, OAM&P can use data path functions or control path functions within the data path. However, if the data path is down or the control path within the data path is down, OAM&P becomes impossible. As a result, OAM&P communication within the router should be carried through dedicated communications channels. It has been proven very cost-effective and more than sufficiently reliable to use Fast Ethernet as a means of communication between the OAM&P entity and the line cards, the switch fabric cards, and in between

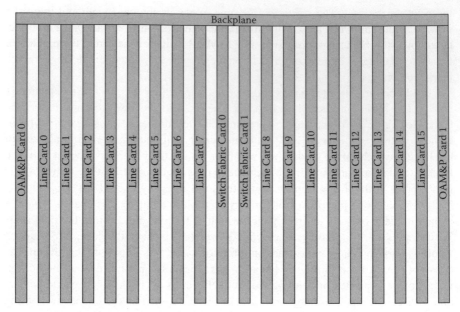

FIGURE 7.10 Backplane and active Cards

potentially redundant OAM&P entities. Communications with fan trays and temperature sensors within the chassis are typically implemented with I²C.

Power Supply Unit

Backplane designs are the workhorses in the data center and in most ISPs. They typically feature redundant power supplies. These may be two power supplies that operate in a load-sharing mode, or more likely three power supplies that are rated such that any two of them can power the router. For the few routers with power consumption below 1600W, local line voltage is the typical input: 110V AC in the US, 220V AC in the rest of the world. Most of the backplane design type routers colocated with Central Office equipment and that need to make use of the Uninterrupted Power Supply into the CO will require a –48V DC power converter. Even if they are not colocated with CO equipment, their power consumption is high enough to require a 220–240V AC power supply. All power supplies must be hot-pluggable. It is very useful for the power supplies to have temperature sensors and fan speed control circuitry. Typically, the power supply would have an I²C bus interface to the OAM&P card.

Fan Trays

A backplane design router would feature two fan trays with six fans each; one below the card cage; the other above the card cage. The fans are set up such that they support the natural thermal convection through the rack. A brief removal of a fan tray to exchange a nonfunctional fan therefore does not significantly impact the temperature inside the chassis. In CO type applications, an air filter is typically installed. The fan trays typically support fan speed control and temperature measurement via an I²C bus interface to the OAM&P card. Fan trays and the embedded fans are hot-pluggable.

CONCLUSION

As we have seen in the previous discussion, the chassis has a significant impact on the router's capabilities and its deployment. Single-board systems are used where cost is an issue, and modular systems are deployed very widely when it comes to high performance and high availability. A wrong choice made in the chassis setup will impact its intended performance and its possible deployment. A single-board design will not perform well in a Central Office environment, no matter what amount of engineering is invested. Conversely, it will be impossible to bring a backplane architecture type router into a price range that would enable it to compete with single-board architectures in SoHo applications.

8 Line Cards

OVERVIEW

The purpose of a router is to forward datagrams from an ingress port to an egress port. During periods of link contention or congestion, it also must make decisions as to when to queue or discard a datagram. The line cards or processor cards in a midplane design derive the forwarding decision, the queuing decision, or the drop decision from information contained in the packets' or cells' header or even in their payload.

Line cards are a category of cards that contains the logical functions for processing data from and contained in all network layers—from the physical layer all the way up to the presentation layer. A line card therefore describes logical functions, not necessarily physical devices. In effect, it connects the physical or logical port containing the Line Specific Logic (LSL) to the switch fabric and ultimately to another line card. A line card contains functions that can be associated with a port or interface card, a processor card, and optionally a traffic management engine, a Segmentation And Reassembly (SAR) engine, and a switch fabric interface. In this scenario, the port or interface card contains the physical interface (PHY)—which might be an electrical or optical interface—the Media Access Controller (MAC), and optionally a framer if the datagrams are in a frame format. The processor card contains a network processor which classifies ingress and optionally egress datagrams, a lookup engine—which might be a CAM or a piece of software on the network processor—to associate destination addresses to port addresses, and the packetizer and depacketizer, or SAR engines. Additionally, the processor card might contain a traffic manager for queuing ingress or egress datagrams. The switch fabric interface may be as simple as a set of parallelized Serializer/Deserializers (SerDes). The optional backpressure signaling from the switch fabric is reported back to the network processor or the traffic manager through the switch fabric interface.

This is significantly more circuitry than in a typical IPv4 router. The reason for the increased complexity is the desired capability to handle all traffic types from TDM and TDM-like traffic to FTP and NNTP traffic, the desire to unify all these traffic types, the increased requirement for robustness and availability, and the contractual agreements in an Service Level Agreement (SLA) between a user and a provider of these services.

LINE CARD FUNCTIONS IN IPv6 ROUTERS

IPv4 line cards have done quite well so far and proven to be more than good enough for cost-efficient "best effort" traffic. However, with the introduction of more advanced services and protocols on top of IPv4, it has become increasingly difficult to maintain any type of Quality of Service (QoS). The mere fact that more protocols and more header information are to be evaluated certainly establishes a challenge all by itself, but it is a fact that IP headers, applications headers in the payload, and some superimposed protocol might contain conflicting information that can make the situation worse.

In fact, an IPv4 router set up with Multi Protocol Label Switching (MPLS) tunnels will face these challenges, depending on how deep the packet investigation goes and if it is enabled at all, or if the router is instructed to only obtain forwarding information from the tunnel setup. As an example, a tunnel might be set up already, and a Label Switch Router (LSR) receives packets to be forwarded. Since the tunnel has been set up, it is clear that there should be capacity in the links available. If the contents of the packets in the tunnel indicate conflicting routing or forwarding information, then the setup of the MPLS LSR will determine whether the tunnel setup information or the packet information prevails. As a result, the packets in the tunnel might have a different priority from what the tunnel setup indicates. Alternatively, internal capacity might not be available if the router must rely on a shared bus or a shared memory switch or a poor design of a switch fabric. In these cases, temporary contention will occur, and the delay variation and maximum latency cannot be maintained. Therefore, the LSR will have to look into the packets, and determine if the packets can be queued, forwarded, or must be dropped. The behavior of the router therefore differs depending on how it is configured. This hopefully ends with IPv6, where the QoS field should be enough to assign a priority to a packet, and no conflicting information is sent. However, this seems more like a goal than reality. In reality, IPv6 routers will have to tunnel IPv4 traffic, and vice versa. Therefore, only an end-to-end IPv6 "connection" will guarantee QoS.

IPv4 routers typically perform a packet investigation by parsing the IPv4 header and all application headers in the payload. This header information then is classified and the data consolidated. Sometimes policing is performed, and then quota and policing tables are updated. Traffic management typically is not performed. Instead, datagrams are directly forwarded to the switch fabric card, or dropped if the egress line is in overload or the policy indicates to discard the datagram.

In contrast, the goal for IPv6 routers is to avoid performing deep packet investigation at all because that will have occurred at the edge or the core edge. That deep packet investigation and consolidation of routing information will have to occur at the edge or the core edge because the IPv6 header will have to contain all pertinent information for it to be useful. In a way, the edge or core edge routers will have to pre-classify the datagrams such that the IPv6 header holds all routing information, including the QoS and path info. If that is the case, then the core IPv6 router can parse only the IPv6 header and rely upon it. There must not be any discrepancies in the routing information at this point in time. With the IPv6 header holding valid and accurate routing information, the IPv6 router can use this info as the sole entity to be parsed, and can make routing decisions. It will police the datagram based on its header information and policing data, and it will perform traffic management functions according to the QoS, priority, traffic management policies, and device and network congestion or load status. Based on this data, the router determines whether any particular datagram will be forwarded to the switch fabric card immediately, enqueued, dequeued, or discarded. These are functions the IPv6 router can perform based on the IPv6 header and all other pertaining information in policy databases. In addition, it must be aware of the load or congestion status of the router and the network in the uplink direction. The advantages of doing this are better network utilization, higher efficiency, and easier management.

It is important to know what these fields mean. In Figure 8.1—the IPv4 header format—we can see the 4-bit version field. It is 4 for IPv4. The next field is the 4-bit Internet Header Length field. It determines the length of the header, and therefore can be seen as the offset to the payload. The following field is an 8-bit field setting the Type of Service (TOS), and is now being used as the DiffServ and Explicit Congestion Notification descriptor. In the field following the TOS is the 16-bit Length field. The length field determines the total length of the packet, including the header and payload. The Identification field is a 16-bit value identifying the sequence number of the packet within a larger datagram. A 3-bit field for Flags determines whether a datagram may be fragmented or not, and if so, if more fragments follow. The following Fragment Offset value is a 13-bit sequence number indicating the sequence number of the packet within the datagram. An 8-bit value called Time To Live (TTL) is used to limit the life (or number of hops) of the packet. Each router the packet traverses decreases the TTL by one, and if TTL reaches zero, the packet is discarded. The Protocol field—8 bit long—describes the protocol type on top of IPv4. The header checksum is a 16-bit field protecting the header. Both Source Address and Destination Address are 32 bit long each. The 32-bit Options and Data fields are rarely used, but are considered part of the IPv4 header.

Figure 8.2 describes the IPv6 header. First, it has the 4-bit Version field. It is 6 for IPv6. The next field is the 4-bit Priority field, describing the priority of the packet. Following the priority field is the Flow Label field, a 24-bit value. Flow Label and Priority together describe the Quality of Service—real-time and packet delay as well as delay variation—requirements of the packet, and its association with a "flow". The flow can best be compared to a virtual channel in ATM. Following this is the 16-bit Payload Length field. It determines the length of the payload. The Next Header field is the IPv6 version of the IPv4 Protocol field. Like the IPv4 Protocol field, it is 8 bits long and describes the protocol type on top of IPv6. An 8-bit

+	0–3	4–7	8–15	16–18	19–31
0	Version	Header Length	Type of Service (now diffServ & ECN)		Total Length
32	Identification			Flags	Fragment Offset
64	Time To Live (TTL)		Protocol		Header Checksum
96	Source Address				
128	Destination Address				
160	Options				
192	Data				

FIGURE 8.1 IPv4 header format.

+	0–3	4–7	8–15	16–23	24–31
0	Version	Priority	Flow Label		
32	Payload Length			Next Header	Hop Limit
64	Source Address				
96					
128					
160					
192	Destination Address				
224					
256					
288					

FIGURE 8.2 IPv6 header format.

value called Hop Limit is used to limit the life (or number of hops) of the packet. Each router the packet traverses decreases the Hop Limit by one, and if Hop Limit reaches zero, the packet is discarded. It is the IPv6 equivalent of the IPv4 TTL field. Both Source Address and Destination Address are 128 bits each.

The ultimate goal is to have the application and the Edge or Core Edge Router assign the QoS parameter based on the users' policy such that the core IPv6 router can focus on forwarding the packets in case there is internal and external bandwidth available, make smart decisions about when to enqueue and dequeue packets, and when to discard packets.

From the introductory chapter we can directly take over the list of requirements. These include the capability of the future router to perform the following:

- Losslessly switch and route datagrams
- Be able to perform Segmentation And Reassembly (SAR)
- Perform policing at line speed for thousands of virtual connections simultaneously and thereby enforce and support SLAs
- Perform Traffic Management by queuing and buffering traffic according to SLAs, and drop excess traffic
- Support Traffic Engineering by means of collecting Statistic Traffic Data
- Support a mixture of hierarchical and mesh interconnect infrastructure in terms of data traffic
- Support an overlay network for metadata and signaling data
- Enable traffic rerouting at the edge and provide redundant fail-safe systems towards the core
- Communicate securely within the components of the router

- Communicate securely between the OAM&P card and a Billing Center
- Communicate securely between the OAM&P card and a PKI Center for authentication
- Communicate securely between the OAM&P card and a Network Management Center (NMC)
- Communicate with the PSTN infrastructure

As a result, we can break these down into categories and derive technical solutions from these requirements. Some of these requirements apply to line cards; others do not. Some requirements that apply to line cards also apply to other components of the router, and some require multiple components to work in conjunction to achieve the goals. Pertaining to the line cards or port and processor cards are the following tasks:

- Losslessly switch and route datagrams
- Be able to perform Segmentation And Reassembly (SAR)
- Perform policing at line speed for thousands of virtual connections simultaneously and thereby enforce and support SLAs
- Perform Traffic Management by queuing and buffering traffic according to SLAs, and drop excess traffic
- Support Traffic Engineering by means of collecting Statistic Traffic Data
- Support a mixture of hierarchical and mesh interconnect infrastructure in terms of data traffic
- Support an overlay network for metadata and signaling data
- Enable traffic rerouting at the edge and provide redundant fail-safe systems towards the core
- Communicate with the PSTN infrastructure

From this list, it becomes fairly obvious that line cards have two distinctly different main functions.

One is to interface to the line or the outside world, on layers 1 and 2 of the International Standardization Organization (ISO) Open Systems Interconnect (OSI) stack. This does not have a direct impact on IPv6, although the net data rate is impacted by the protocol because of the fixed line baud rate and the protocol overhead consuming net payload bandwidth. Processing of layers 1 and 2 is performed by a Physical Layer IC (PHY) and a Media Access Controller (MAC). A framer might be required for Packet over SONET (PoS) applications as well, but it is not necessary in each case.

The second part is the higher layer processing; here is a direct impact of IPv6. IPv6 has not only a longer header than IPv4—and therefore requires larger lookup tables—it also has additional features like geographic addressing and QoS parameters. In addition, tunnels for MPLS and IPv4 must be detected and handled appropriately; more importantly, connection setup and connection teardown must be processed for TDM-like traffic.

Even more importantly, the higher layer processing must take care of QoS and latency parameters of the datagram. It must determine all relevant information pertaining to the datagram, and then act accordingly. It must determine the priority of the datagram, its maximum latency allowed, and the requirements for the maximum deviations from the Time Division Multiplex (TDM) timeslots of the Public Switched Telephony Network (PSTN). Once that is done, the datagram either must be forwarded immediately to the switch fabric card, it may have to be enqueued and dequeued in the egress-side Traffic Manager in case the egress port is congested, or it must be discarded because there is no possibility of fulfilling the SLA requirements under current circumstances.

Since all multiprotocol-capable services such as ATM and IPv6 define a Quality of Service (QoS) field, different services and traffic types can be routed through the same router. However, the router (and for the most part, the line card) must distinguish between the traffic types and treat them according to the SLAs' parameters. This has been tried multiple times before—with ISDN and ATM. Both technologies did not really ever take off. However, unification of the communications infrastructure is in full swing right now. One of the reasons is that now computer power is sufficiently cheap that the different traffic types can be tagged appropriately, and that traffic can be routed and forwarded according to its QoS parameters. This ultimately can be used to unify all the traffic types that are in use today onto a single network infrastructure. Where ATM defined Constant Bit Rate (CBR), Variable Bit Rate/Real Time (VBR-RT), Variable Bit Rate/Non-Real Time (VBR-NRT), Available Bit Rate (ABR), and Unspecified Bit Rate (UBR), IPv6 defines 16 priorities that can be used to implement several different levels of QoS. 8 of those are reserved for traffic that backs off (non-real time traffic), and 8 for traffic that does not back off (real-time traffic).

However, IPv6 is not as sophisticated as ATM because it does not have to be. Where ATM defined all the above parameters for the variance of the bit rate and the cell delay variation (CDV), IPv6 defines only levels of priority for QoS, and this is ultimately what the routers and their line cards and switch fabrics can understand and process. While it is nice to have CBR, VBR-RT, and VBR-NRT as available traffic types, it makes the design of a line card extremely complicated because arrival and departure of cells must be policed and monitored. It is computationally much more useful to reduce the complexity down to priorities with which cells are processed and forwarded from the ingress line card through the switch fabric to the egress line card. In effect, IPv6 provides the same or better throughput at the same or better compliance with prioritization as ATM, but it does not provide guarantees for real-time transmittals of data. There is no provision for measuring or influencing CDV or packet delay variation in IPv6.

The parameters that impact the SLAs are under at least partial control of the line card. We can conclude that the line cards will have to control and maintain the following parameters directly in order for the routers to be able to comply with the SLAs' requirements:

- Net bit rate (minimum, maximum, average)
- System availability (minimum)
- System uptime (minimum)
- Cell Delay Variation (CDV) (minimum, maximum, average) (ATM networks only)
- Logical Connection setup time (minimum, maximum, average)

- Delay and Latency (minimum, maximum, average)
- Round-Trip Delay (minimum, maximum, average)

As a result, these routers will be able to effectively route and switch multi-protocol traffic of different types.

DEFINITIONS

DEFINITION OF A LINE CARD

A line card is a subsystem that receives datagrams on an external ingress or internal link from the switch fabric card. It processes one or more datagrams to be able to make forwarding, queuing and dequeuing or discarding decisions, and then executes these decisions. It does this by parsing, processing, and policing the datagram or a set thereof according to parameters derived directly or indirectly from a database. Accordingly, it either discards or sends out the datagrams to the external egress link or the internal link. If the datagram size and type used externally differs from the internally used datagram size and type, the line card will have to perform Segmentation And Reassembly (SAR). In general, a line card processes datagrams on the ISO OSI layers 1–4, with the layers being layer 1 (physical), layer 2 (data link), layer 3 (network), and layer 4 (transport). However, in order to extract the required routing And forwarding information, the line card might have to be configured such that it parses and processes information from all layers above 3 on the ISO OSI model. In order to do so, it will remove and add a variety of wrapper layers or encapsulation layers.

DEFINITION OF A PORT CARD

A port card is a subsystem that receives datagrams on an external ingress or internal egress link and processes one or more datagrams. It sends out the datagrams to the external egress link or the internal link. In general, a port card processes datagrams on the ISO OSI layers 1 (physical) and 2 (data link). In order to do so, it will remove and add a variety of wrapper layers or encapsulation layers. The port or interface card connects the external line physically and logically to the processor card. On ingress, it removes all line coding, frame structures, and unnecessary headers, while on egress it adds the new headers, the line coding, and frame structures. In short, it contains the line specific logic (LSL) and deals with the LSL protocols.

DEFINITION OF A PROCESSOR CARD

A processor card is a subsystem that receives datagrams on an internal ingress link from the port card or an internal link from the switch fabric card. It processes one or more datagrams to be able to make forwarding, queuing and dequeuing, or discarding decisions, and then executes these decisions. It does this by parsing, processing, and policing the datagram or a set thereof according to parameters derived directly or indirectly from a database. Accordingly, it either discards or sends out the datagrams to the external egress link towards the port card or the internal link. If the datagram size and type used externally differs from the internally used datagram size and type, the line card will have to perform SAR. In general, a processor card processes datagrams on the ISO OSI layers 3 (network) and 4 (transport). However, the processor card might have to be configured such that it parses and processes information from all layers above 3 on the ISO OSI model in order to extract the required routing and forwarding information. While doing so, it extracts essential data out of the datagram to provide routing and forwarding information. In order to do so, it will remove and add a variety of wrapper layers or encapsulation layers.

FUNCTIONAL REQUIREMENTS

Requirement 1: losslessly switch and route datagrams. An important item on the list is the ability to losslessly switch and route datagrams. This must occur in conjunction between the line card and the switch fabric card.

The line card parses ingress datagrams, extracts their essential routing and forwarding information, segments the datagrams and prepends the tag, and then forwards it to the switch fabric card—which ultimately forwards the datagrams to the egress-side line card. While the line card performs functions on layer 1 (physical), layer 2 (data link), layer 3 (network), and layer 4 (transport), the switch fabric card provides the logical and physical connections between line cards. However, the more compute-intensive problem lies on the side of the line card. In order to be able to determine the routing and forwarding information, the line card must remove all wrappers of the ingress datagram and possibly even reassemble commands, messages, data, or status information, maybe even within the payload of the datagrams. In the case of UML, XML, HyperText Markup Language (HTML) "switching," or load balancing over IPv4, IPv6, and ATM, some information that is crucial to making the forwarding decision is actually contained in higher layers of the ISO OSI model, and not in the IPv4, IPv6, or ATM header; some information is not contained in the headers or the payload at all, but is based on status or backpressure information the router receives through other means. As a result, the line card must be able to understand the part of the payload that might carry that information. In order for the switch fabric to losslessly switch and route datagrams, the line card will have to generate a locally understandable tag such that the switch fabric card can act upon the forwarding information (a Local Connection Identifier, LCI). It will have to contain at least the egress port number and the priority at which that particular datagram will have to be forwarded through the switch fabric card—even if the line card will have to segment the datagram on ingress and reassemble it on egress. If SAR is required, the LCI will also have to contain the originating ingress port number and a cell sequence number to ensure correct reassembly of the externally used datagram.

The effort the processor card or the processor part of the line card will have to spend to determine egress port numbers and priorities (LCI) depends on a variety of parameters. One parameter is the depth of investigation that is required to derive the LCI from the datagram. It also depends on the line rate, the type of switching required (circuit switching = quasi-static; packet or cell switching = dynamic), the cell or packet length, and therefore the arrival rate of the packets or cells. This is fairly basic, but crucial to the operation of the line card. While in a circuit switch, a connection—even a virtual connection—is set up at the beginning of the connection or transmittal (physical, logical, or virtual) and torn down at the end of it. A packet or cell switch does not set up or tear down connections, but rather forwards every packet or cell individually based on the routing information contained in each and every packet or cell. Therefore, a packet or cell switch is fundamentally more compute-intensive. While it offers a higher flexibility and a potentially better utilization of the line, it requires much more compute power to determine the LCI from information contained in the datagrams—all at the arrival rate of packets or cells. The arrival rate of cells or packets therefore determines the rate at which the line card (or the processor card, for that matter) will have to be able to find the appropriate tags in the datagrams, look up essential routing and forwarding information, and then apply this as the LCI to the segmented internal datagram to forward it to the switch fabric card, enqueue it if required, dequeue it if possible, or discard it if necessary. As a result of this dynamic switching and forwarding of packets, there is no Connection Admission Control (CAC), and therefore no predetermined departure rate of packets or cells on a per-egress link basis. A consequence of this is that neither the ingress-side network processor nor the egress-side network processor can avoid egress line contention. By definition, contention of the egress line is not only a possibility, it is a fact of life in packet or cell switches. While mostly this contention is of very short duration, it can span longer periods of time. Statistically, the distribution of packet or cell destination addresses and priorities should follow a stochastic distribution, and that limits the extent and the duration of contention on the egress line. However, not all traffic is equally distributed, and some event within the network may cause a significant deviation thereof. In any case, assumptions can be made as to what the distribution is and how it affects the egress line. Dealing with this problem is the task of the traffic manager, and not the issue of the network processor on the line card. Shaping or managing the traffic has indirect impact on the lossless switching and routing of datagrams, and is described in more detail later. Lossless switching and routing in this context means that datagrams are switched and routed such that there is no unintentional datagram loss. Intentional loss of datagrams results out of decisions made by the network processor and the traffic manager based on egress link congestion and contention—as well as on quotas derived from policies—and affects low-priority timing-insensitive datagrams first, then escalates until the overload situation is resolved. Not instituting a drop policy likely will result in larger proportional loss of datagrams, which will violate QoS and SLA parameters, especially for high-priority timing-sensitive data.

The IPv6 router's line card parses all datagrams at the arrival rate of packets or cells. This is not new. What is new is the fact that the header is longer than in IPv4 routers. Larger Source Address (SA) and Destination Address (DA) fields in IPv6 compared to IPv4 were a fundamental requirement of the study group. It was becoming clear that there was and would be an address shortage, especially considering that the classification in Class A, B, and C subnets made the use of the existing addresses even less efficient. Therefore, it was clear that IPv6 would provide longer address fields. The IEEE opted for 128 bit SA and DA fields, compared to the 32 bit each in IPv4. The address fields are four times as long. To many observers, the increase seems like a fourfold capacity, but the address space is not only four times as big. We will see the impact of the length of the SA and DA fields.

The address space is extended from 2^{32} bit to 2^{128} bit, and $2^{128} = 2^{32+96} = 2^{32} * 2^{96}$. Therefore, the address space IPv6 offers is 2^{96} times as big. The extension of the address fields actually provide a 2^{96}-fold capacity—which appears to be a good thing, since it will provide enough address space for the rest of the life of the Internet Protocol. It does, but to achieve this, a 64 bit entry would have been sufficient—this would have been a 2^{32}-fold increase in address space. One may ask what is wrong with an address space that is too large. Theoretically, nothing is wrong with it other than that the address space will be sparsely populated. In theory, this is something that can easily be handled. In reality, one must remember that these entries in a Content Addressable Memory (CAM) must somehow be physically implemented. Even considering a lookup that depends only on SA, DA, and QoS with a three- or four-bit-wide field, it is based on a 260-bit value. 2^{260} bit represents a huge address space, and requires huge lookup tables if fully populated. If a six-transistor cell is used for each bit, adding 15% overhead for CAMs over an SRAM, the total is $1.15 * 2^{260}$ entries, or approximately $1.15 * 10^{85}$ entries. If we can make a transistor out of 100 atoms, we need $1.15 * 10^{87}$ atoms to build our CAM. Unfortunately, our universe only contains about 10^{80} particles—baryons and leptons. Therefore, it would take more material than our universe contains to build a CAM that allows mapping all SA/DA/QoS combinations to a route and a priority. It just is not practical and not possible. If it is not possible, then the question is—why was it done? For the most part, the engineers suggesting the scheme wanted to escape the problem of scarce address space once and for all. Logically, they managed to do so, but in reality, 64-bit fields would have worked. Therefore, it is important to understand that, if the CAM is seen as a consecutive set of table entries, these huge address spaces are logical address spaces with lots of empty spaces between entries. In reality, the limited number of egress ports, multiplied by the number of priorities the router is supposed to support, will determine the number of entries in the CAM anyway. A 32-port router with 8 priorities will have to map all IPv6 SA/DA/QoS tuples to one of 256 possible paths. Therefore, a CAM that supports 256 destinations mapped to a larger number of SA/DA/QoS tuples is sufficient, and the vast majority of the address space goes unused because it is mapped into a very small number of physical or logical paths. The only difference is that in an IPv6 router the input parameters are even more sparsely populated than in an IPv4 router, and there are even fewer consecutive islands of addresses that are even smaller in size than in an IPv4 router. As a consequence, entries are more scattered and scarce than would be desirable for a ternary CAM. That has an even higher impact on software-based lookups than the impact it has on CAMs. Whereas a CAM does not lose throughput based on scattered data, a hashing algorithm on a general-purpose CPU does. The CAM will only grow in size to implement scattered data, and therefore will be more expensive and will require more power, but it will not sacrifice throughput. A general purpose CPU will have to perform more compares and branches if data is more scattered, and therefore lose throughput in terms of lookups per second with smaller islands of entries and more scattered data. The additional problem faced by the software-based database on a general purpose CPU—even if we could make one that holds translation data for 260-bit SA/DA/QoS fields—is that it will require 1 to 260 accesses to find the entry. On average, it will take 129 accesses. If each access takes one CPU cycle—which is too optimistic of an assumption—it will take 129 CPU cycles to find the entry with an interval bisection method. To speed it up by a factor of two, a better algorithm will have to be devised. This better algorithm will have to use no more than one cycle per access in order to be more effective. A parallel lookup in a CAM is—as we have seen—not feasible because of the size of the CAM required. A segmented CAM that supports only a small subset of all potential addresses will perform the lookup sufficiently fast, and will not exceed reasonable memory sizes and therefore die sizes.

If too many entries in the routing table are unpopulated or point to the default forwarder, then the routing efficiency is reduced, and too many packets will have to be treated with a default priority—which in turn either consumes too much of the more expensive high-priority bandwidth or violates the policies for high-priority packets and, therefore, does not fulfill the QoS guarantees in the SLA.

In other words, the sheer size of the logical address space constitutes a huge challenge. A fully populated lookup table of this size is not feasible, and even if it were, both a sequential lookup from a software-based lookup engine in a general purpose CPU and a segmented (partially sequential) CAM would have a fundamental problem looking up the routing parameters in an acceptable time frame—even if the search is only partially sequential. Therefore, the lookup table must be reduced to a practical size. However, dealing with sparsely populated routing tables poses a challenge all by itself. Retrieval is the easier part, but maintaining a sparsely populated routing table, a routing table with non-consecutive entries, or even one with entries that are out of sequence, is difficult. Retrieval of the data therefore is not the only challenge faced by a CAM or a software-based lookup engine. The initial task is that the table will have to be configured. Second, the table must be maintained. The larger it is, the more likely that its entries change during any given period of time, and therefore routing table updates become more frequent with larger numbers of entries. Routing table entries age and consequently must be updated. As a result, the growing size of the number of table entries will constitute a massive growth in required signaling traffic, and thus reduce net payload efficiency. Additionally, the CAM or the software-based lookup engine will spend proportionally more time updating the routing table. During these updates, either read access will be blocked and therefore routing table information cannot be retrieved, or the read access will render an invalidated entry.

We have established that the routing tables are required to determine essential routing information. The routing table size required for IPv6 makes the routing table and processing thereof a non-trivial part of the router. We additionally must determine at which rate we will be required to retrieve routing table entries to allow line rate datagram processing, and therefore datagram routing and forwarding.

In order to compare the requirements that result out of line rate per-datagram lookup and processing, Table 8.1 gives an overview of the approximate number of datagrams per second for typical line cards.

TABLE 8.1
Datagrams per Second on Select Communication Link Technologies

Technology	Datagrams per Second		
	Line Rate in bit/s	Datagram Length in bytes	Datagrams per Second
Cell Phone	9600	1	1200
PSTN	64000	1	8000
56k Modem (PPP)	56000	64	109
T1	1544000	1	193000
E1	2048000	1	256000
PPP over T1	1544000	64	3016
PPP over E1	2048000	64	4000
10 Mbit/s Ethernet	10000000	64	19531
ATM STM-1/OC-3	155000000	53	365566
100 Mbit/s Ethernet	100000000	64	195313
ATM STM-4/OC-12	622000000	53	1466981
1 Gbit/s Ethernet	1000000000	64	1953125
ATM STM-16/OC-48	2488000000	53	5867925
10 Gbit/s Ethernet	10000000000	64	19531250

Table 8.2 describes the approximate number of MIPS required to perform the functions of a typical line card, dependent on the line rate. We assume that any of the above-mentioned operations—including parsing the datagrams, a lookup, and processing the datagrams—take 100 CPU cycles. These numbers are intentionally significantly underestimated. The additional columns show the MIPS requirements for 200, 500, and 1000 CPU cycles per operation.

As a comparison, the largest PSTN Class 5 CO switches (Elecktronische Waehl-System Digital, or EWSD, from Siemens Public Networks) can handle 1M Busy Hour Call Attempts (BHCA). This includes connection setup, connection teardown, and making an entry into the billing table once every 15 minutes. In other words, it can handle 1 million call setup messages (and set up the connections), 1 million call teardown messages (and tear down the connections), and 4 million billing entry updates, all within 3600 seconds. That translates to 2 million messages parsed and processed, and 4 million billing table entry updates. Assuming that each operation requires 100 CPU cycles, EWSD provides 600 million CPU cycles within 3600 seconds, or 166667 CPU cycles per second for call processing, policing, and billing—not counting other potentially internal activities. The total aggregate bandwidth of that particular EWSD switch is 300,000 users with 64 kbit/s each, or 19.2 Gbit/s. It is quite amazing that dealing with 300,000 users and a total capacity of nearly 20 Gbit/s requires so little effort. In other words, circuit switching allows for some quite leisurely processing of data.

While quite a few CPU cycles within EWSD (or any other Class 4 or 5 Central Office Switch, for that matter) are used to ensure failsafe operation, the requirements

TABLE 8.2
MIPS Required for Select Communication Link Technologies

| | MIPS versus Line Rates | | | |
Technology	MIPS (100 cycles)	MIPS (200 cycles)	MIPS (500 cycles)	MIPS (1000 cycles)
Cell Phone	120000	240000	600000	1200000
PSTN	800000	1600000	4000000	8000000
56k Modem (PPP)	10938	21875	54688	109375
T1	19300000	38600000	96500000	193000000
E1	25600000	51200000	128000000	256000000
PPP over T1	301563	603125	1507813	3015625
PPP over E1	400000	800000	2000000	4000000
10 Mbit/s Ethernet	1953125	3906250	9765625	19531250
ATM STM-1/OC-3	36556604	73113208	182783019	365566038
100 Mbit/s Ethernet	19531250	39062500	97656250	195312500
ATM STM-4/OC-12	146698113	293396226	733490566	1466981132
1 Gbit/s Ethernet	195312500	390625000	976562500	1953125000
ATM STM-16/OC-48	586792453	1173584906	2933962264	5867924528
10 Gbit/s Ethernet	1953125000	3906250000	9765625000	19531250000

are fairly low. The common Fast Ethernet router must process nearly 200,000 packets (see Table 8.2) per second per line card. Its lookup engine must be significantly faster than all of EWSD together. We can see that packet or cell switching is significantly more compute-intensive than circuit switching. The numbers for 10 Gbit Ethernet really pose a problem for a line card. 10 Gbit/s Ethernet requires a sustainable parsing and processing rate as well as a lookup rate of nearly 20 million lookups per second. That rate significantly limits the options of how to process it. In a database of 1M entries, interval bisection takes a maximum of 20 accesses to find an entry ($2^{20} = 1$M); on average, 10 accesses. Even if an access takes only one CPU cycle, the CPU clock frequency must be at least 400 MHz (20 million lookups at 20 accesses per lookup each) to be able to find the correct entry for routing and forwarding the datagram, since we must handle the worst case numbers. In reality, the number of cycles is closer to 10 CPU cycles per access, requiring a 4 GHz CPU. This in turn forces the designer of the router's line card to provide additional resources on the line card for performing the rest of the functions in hardware or within another CPU. Additionally, the memory subsystem must be capable of sustainable 400M read accesses per second for lookups only—not even counting write accesses for simultaneous updates of the table.

On top of that, the power budget for a router is strictly limited. The typical 16-port router cannot consume more than 3500W, which in essence means that no card can consume more than 200W. While this is the absolute maximum limit, most router manufacturers limit the power consumption to 15W on the processor card or the processor part of the line card. As an example, we will determine the values for a Fast Ethernet router, and assume it takes 100 or 500 cycles for each operation mentioned above. It is important to stress one more time that these numbers are sustainable throughput numbers just for task processing, and not peak performance MIPS or MIPS/W, excluding all overhead, Interrupt processing, task synchronization, task switching, and cache misses. Adding up all required MIPS for the network processor tasks, the traffic manager tasks, and the policing tasks in both directions on the line card according to Table 8.2, and dividing the numbers by 15W gives us $3 * 2 * 20$ MIPS $= 120$ MIPS required. Therefore 120 MIPS/15 W $= 8$ MIPS/W is the first target number for the compute efficiency we will have to request when selecting a CPU and the appropriate memory subsystem that performs parsing, lookup, and traffic management, and executes the forwarding, queuing and drop decisions. Another measure is that most router manufacturers limit the device power consumption to 600W total for 12 line cards, one switch fabric card, and one OAM&P card, therefore allowing approximately 40W per card. This leaves about 15W for the processor card or processors part of the line card. As a comparison, a 3GHz Pentium 4 delivers significantly less than 1000 MIPS of application performance at a power consumption of around 100W, and therefore delivers significantly less than 10 MIPS/W.

Requirement 2: perform Segmentation And Reassembly. SAR is a very fundamental and important function on a line card. As we have seen earlier, routers on their port cards predominantly receive packets on their ingress ports and send packets out of their egress ports. However, the switch fabric card delivers by far the best throughput with fixed-size datagrams, also called cells. Therefore, on the internal links, the line card will have to use cells. As a result, all advanced routers will have to segment packets on the ingress side from the port cards into cells to be forwarded through the switch fabric card, and then reassemble the cells into packets on the egress side. This is not required for packets that terminate in the ingress-side processor card (EGP, RIP, other IPv4/v6 signaling, or other metadata such as queue status information), but all other packets that must be forwarded through the switch fabric will have to undergo SAR. Even packets that terminate in the egress-side processor card will have to be segmented on ingress and reassembled on egress.

In theory, SAR is a somewhat trivial operation: datagrams of variable or constant length are converted into fixed-size cells of a preset length. The incoming datagrams are segmented into fixed-size internal datagrams, and a Local Connection Identifier (LCI) is prepended to each of these internal datagrams. The LCI must contain at least the destination port address. In practically all cases in a modern router, it will have to contain the priority of the datagram, any other descriptor for its requirements regarding real-time forwarding, plus aging descriptors. If Reassembly on egress is required and no other means of identifying the ingress port are provided, then the LCI will have to contain the ingress port number. If there is any possibility that internal datagrams arrive at the egress port side line card and its Reassembly engine is out of sequence, then the LCI will also have to contain a Cell Sequence Number (CSN) with a range that is greater than or equal to the integer part of the quotient of the largest allowed external datagram size divided by the payload size of the internal datagram. This is necessary to identify internal datagrams out of sequence, and possibly even repair out-of-sequence arrivals of internal datagrams. If cells never can arrive out of sequence, then the LCI will require a flag indicating that the last cell for the packet has arrived. These requirements appear rather simple and seem not to pose a problem.

However, the details of SAR—especially when combined with requirements for minimizing packet or cell delay and packet or cell delay variation—are less trivial. On the Segmentation side, the incoming datagrams are segmented into internal cells of constant length and the LCI is prepended to the front of the internal cell. The resulting delay and the delay variations depend only on the datagram arrival rate. The LCI allows the switch fabric card to forward the cell to the desired egress port at the determined priority. Once the cell has arrived at the egress port, it will await Reassembly. This, however, is not a simple task. First of all, cells arriving at any egress line card can come from any ingress line card. They may also have different priorities assigned to them, and more importantly, may or may not be interleaved. While the sequence is maintained for each class of cells on a per-priority, per-ingress port basis, this does not mean that it is maintained for all cells on a per-priority or per-ingress port basis. As a result, the Reassembly buffer on the egress line card must accommodate at least enough buffer entries for all ingress ports multiplied by the number of priorities supported by the router. Ideally, the number of Reassembly buffer entries is double that number. This allows for packets or cells to be reassembled even if one internal cell is lost and therefore the previous packet or cell cannot be completed. Instead of discarding the incomplete packet or cell and the current one, only the previous incomplete packet or cell must be discarded. As a result, the number of Reassembly buffer entries ("Reassembly buckets") can be quite high. In a router with 16 ports and 8 priorities, the minimum number of Reassembly buckets is 128 on each line card, and the optimum number is 256. If the router has 16 physical ports that provide 256 virtual or logical ports, and each of them supports 8 priorities, the number of Reassembly buckets must be 2048 (minimum) or 4096 (ideally), which makes it harder to implement in hardware. Instead, a software solution that can perform pointer processing at reasonable speeds may be the preferred solution here. This does not yet solve the problem of latency, delay variation, and determination of packet or cell completeness. Obviously, the latency of the reassembly process depends heavily on the internal CDV of the switch fabric card and the distribution of high-priority versus low-priority internal cells. It also depends heavily on the scheduling algorithm for prioritization in the switch fabric card. If high-priority internal cells are always favored, then packets or cells comprised of low priority internal cells cannot be reassembled on egress, and the completion stalls. The Reassembly engine must assign a limit on the time for which it tries to attempt reassembly for each and every external datagram. If this time is exceeded, a timeout will lead to abortion of the reassembly attempt. In other words, the scheduling algorithm in the switch fabric card determines the rate at which packet or cell completion stalls in the egress-side line cards' SAR engine. In order to be able to abort reassembly processes of low priority packets or cells, it must be known to the line card what scheduling algorithm in the switch fabric is used. If the line card aborts them too early, the remaining incoming cells will not complete the reassembly of previously aborted packets or cells; they will add to more incomplete packets or cells on egress because the remaining incoming cells cannot be associated with any packet or cell reassembly currently in progress. If a line card aborts the reassembly attempts too late, then the number of Reassembly buckets will have to be increased, and the latency of packets or cells being reassembled will increase dramatically. It will additionally increase the packet or CDV of datagrams going through the router. This is true especially for lower-priority packets or cells. Both effects are highly undesirable. The specifics of pointer processing or any other implementation of the Reassembly process in the SAR have a comparably small effect on the delay, the delay variation, or the completion rate.

As we can see in Figure 8.3, the segmentation engine will first have to fill up ("pad") the packet to the next integer multiple of the internal datagram payload size (segments "S"), then cut the resulting entity into segments of the internal datagram payload length, prepend the LCI, and send it on its way to the switch fabric. The switch fabric forwards cells based on the LCI only.

Figure 8.4 illustrates the reassembly process. All segments that belong to the same packet are transferred into consecutive memory locations while the LCI is stripped off. The datagram that is read out of the consecutive memory locations then is the original datagram with the additional padding in front of its header. After removal of the padding, the packet is in its original form. The reassembly engine

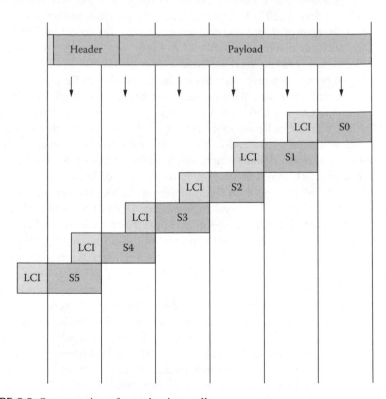

FIGURE 8.3 Segmentation of a packet into cells.

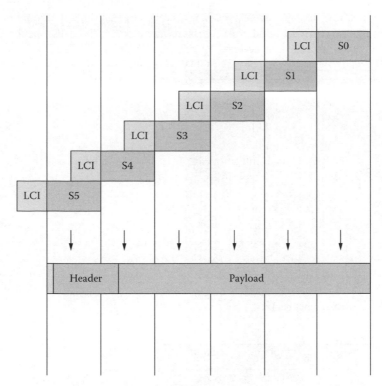

FIGURE 8.4 Reassembly of cells into packets.

ultimately determines the packets' composition from its LCI. Assume no reordering of subsequent cells can occur during their traversal through the switch fabric. If cells can arrive out of sequence, then the Cell Sequence Number (CSN) is used for reordering the cells upon arrival in the reassembly engine. In Figure 8.4, it is assumed that all incoming cells are originating from the same ingress port and have the same priority. That is not a realistic case; the realistic case is illustrated in Figure 8.5. The CSN can help determine when to abort a reassembly attempt and when to reorder out-of-order cells, as can be seen in Figure 8.6.

Figure 8.5 shows that the incoming stream of cells into the Reassembly engine interleaves cells from different ingress ports at different priorities. These incoming cells must be transferred into buckets of the appropriate tuple of ingress port number and priority for reassembly into packets. The example shows an 8-port router with 4 priorities for each of the physical and logical ports. As a result, the router has 32 reassembly buckets into which incoming cells must be transferred in the sequence of their arrival. Again we assume that no reordering of subsequent cells can occur during their traversal through the switch fabric. If cells can arrive out of sequence, then we must use the CSN for reordering the cells upon arrival in the reassembly engine.

Even in case the external datagram type is a cell—not a packet—with a size different from the internally used cell size, the Segmentation And Reassembly process ("SARing") can be quite a challenging process.

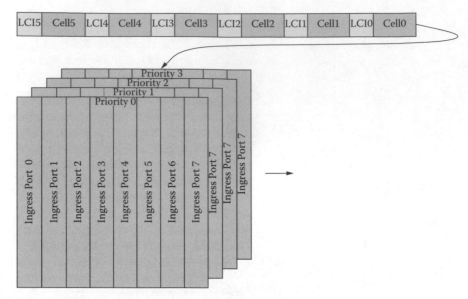

FIGURE 8.5 Reassembly buckets.

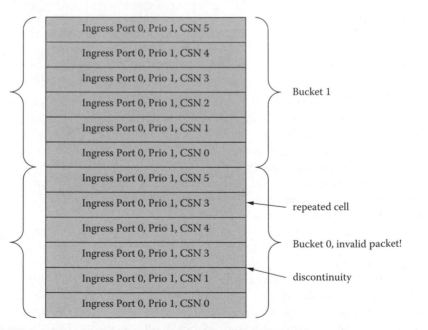

FIGURE 8.6 Packet reassembly with dual reassembly buckets.

SARing requires some processing power and therefore resources; compared to a wire, it contributes to additional cell or packet delay and cell or packet delay variation. It decreases the efficiency of links and of any other transport medium.

That is what opponents of ATM (and any other cell-based switching technology) from the beginning on have dubbed the "cell tax." The idea of the "cell tax" was brought up early in the dispute between ATM and packet networks, such as IP, to discredit ATM and any other cell-switching technology. While it is true that SARing takes time and adds to the latency of datagram traversal through a switch or a router compared to just a bare wire, the combined effect has little impact. More importantly, the fact that a packet switch is significantly less efficient than a cell switch and the fact that its scheduler is significantly more complex leads to the preference of cell switches even in routers—which inherently forward packets. Look at the "cell tax" argument in more detail to understand why it is valid and invalid at the same time. Segmenting variable-size datagrams results in inefficiencies. The worst case is always a packet that is $n+1$ bytes long when the internal payload of the internal cells is n bytes. Obviously, it will take two internal cells to transport the external datagram from the ingress port to the egress port. The first cell will transport n bytes, and the second will transport 1 byte with $n-1$ bytes unused. There will always be waste of bandwidth for as long as the packet size is not an integer multiple of the internal datagram payload size.

The number of cell-sized fragments N_C and the number of fragments N_F in size between 1 and the internal cell size minus 1 both are a function of the internal cell size L_C and the largest allowable packet size L_P.

From this, we can derive that the internal cell size impacts the efficiency of the line and the switch fabric significantly. The distribution of the internal size of the internal cells to be forward through the switch fabric depends on the fragments length L_F ranging from 1 to the internal cell size minus 1 and the number of the fragments having the internal cell size.

If $L_P = L_F + n * L_C$, with $L_F = 0$ or L_P mod $L_C = 0$, then the number of fragments N_F sized from 0 to the internal cell size minus 1 is the modulo of the largest allowable packet isze L_P and the internal cell size L_C each, or $N_F = L_P$ div L_C, and the number of fragments having the cell size $N_{F,C}$ is given by $N_{F,C} = \frac{1}{2} * L_P$ div $L_C * (L_P$ div $L_C + 1)$. If $L_P = L_F + n * L_C$, with $L_F \neq 0$ or L_P mod $L_C \neq 0$, then the number of fragments N_F sized from 0 to L_P mod L_C is the modulo of the largest allowable packet size L_P and the internal cell size L_C each, or $N_F = L_P$ div $L_C + 1$, and the number of fragments N_F sized from L_P mod $L_C + 1$ to the internal cell size minus 1 is the modulo of the largest allowable packet size L_P and the internal cell size L_C each, or $N_F = L_P$ div L_C, and the number of fragments having the cell size $N_{F,C}$ is given by $N_{F,C} = \frac{1}{2} * L_P$ div $L_C * (L_P$ div $L_C + 1)$.

In any case, for all packet lengths greater than twice the cell size the number of cell-sized fragments is larger than the number of fragments for any other size. As a result, in that case the efficiency of a cell switch is greater than 0.5 under all conditions.

The realistic case is that the two maxima are at 64 bytes for all the HTTP put commands and 512 bytes packet length for any file transfers, and taper off very quickly after that. As a result, the efficiency is skewed even more towards 1 because of a further predominance of the 64-byte fragments and cells. We can therefore assume that the efficiency of a cell-based line card and switch fabric is greater than 0.5 and less than 1.0–but it is closer to 1.0 than it is to 0.5.

This is a huge improvement over packet-based line cards and switch fabrics, where the efficiency becomes smaller when the range of the distribution of the packet of fragment sizes get larger. The efficiency of these packet-based line cards and crossbar switches is significantly below 0.5–if not below 0.10. The same is true for DRAM usage if fixed sized memory structures are used for Reassembly. This is another advantage of pointer processing with variable datagram sizes.

However, the convergence of n to infinity is not a realistic case because in reality n does not converge to infinity. One could argue that making the cell shorter increases the efficiency. This would be true if no LCI were required. A 1-byte cell length would of course accommodate all datagram sizes, and would never face the $n+1$ cell problem. However, the LCI is of a certain length that is not zero. Mostly, it is in the range of 8–12 bytes. As a result, a 1-byte payload cell would result in an efficiency of 6–11%. This is not reasonable. The optimum cell size therefore is a compromise between LCI overhead and wasted bandwidth due to overlong payload cells.

Realistically, most packets are small, and therefore a 64-byte net payload cell size is a very reasonable size. The effect of lost time slots is much more pronounced because of switching variable-length packets. Not only is the scheduler significantly more complex, the crossbar cannot switch on fixed-length intervals. It must switch based on the maximum length of a packet in transit through the crossbar, and renders all other timeslots useless, even if all other packets were small and have finished traversing the crossbar. Consequently, the scheduler will have to track the traversal times of packets through the crossbar and schedule connections based thereupon. This fact leads to inefficiencies of packet switches that are so significant that the disadvantages of SARing are more than outweighed.

As we have seen, SARing unfortunately adds to the delay and the delay variation of datagrams traversing the router and it uses up quite a few valuable resources. To minimize the adverse effects, combining SAR with traffic management on egress would therefore be very useful because the reassembly engine can reuse memory from the traffic management engine; at the same time it is in better control of the datagram delay and its variation. Additionally, the datagram delay variation caused by reassembly can be controlled better than if those engines were separate and reassembly occurs after traffic management is performed.

Requirement 3: perform policing at line speed for thousands of virtual connections simultaneously and thereby enforce and support SLAs. In case the router must apply policies to routing the datagrams, it must look up vital routing information for each and every datagram, apply the policies, and appropriately generate an LCI tag that reflects these policies. Very often, the egress-side line card then must check whether the policies were obeyed and generate an entry in the database to reflect the compliance or non-compliance with the preset policies. In any case, both the ingress and the egress line cards record the datagram, its arrival time, its departure time, and the time elapsed. Both ingress and egress-side line cards will have to record the datagram and make a billing or policy database entry after the datagram has been associated with a policy. Therefore, both ingress and egress line cards police traffic. Again, excessive growth of the tables for policies and for billing entries will be detrimental because of the number of tables involved and the number of subsequent lookups. Policing every single packet also allows for checking compliance with the policies and the quotas. This allows for deriving and executing a drop decision upon exceeding the quotas or violating a policy. The fundamental side effect and advantage of this check upon quotas is that drop decisions of packets can be executed immediately. As a result, if there is a traffic manager on ingress, it makes sense to combine policing on ingress with traffic management and shaping into the switch fabric. If there is no traffic manager on ingress, then the network processor will have to perform the policy lookup and execute the decisions, including the drop decision.

Policing requires parsing of each and every datagram and determining its forwarding information from a policy database (see Figure 8.7). It additionally relies on collecting statistical data and histogramming on a per-Source Address and per-QoS basis (see Figure 8.8), based on the information the network processor derives from the IP packet. In some cases, it may be the IPv6 Flow ID bit, and therefore will assign a specific tag in the LCI. As a result, for each incoming datagram a simple lookup will return the number of packets that have been forwarded since a particular point in time. A comparison with the policies set forth in the policy database will then determine if the quota is exceeded or not. If the quota is not exceeded, the datagram will be forwarded according to the policy. If the quota is exceeded, then the datagram will be dropped. There are multiple ways that the drop decision can be derived and executed. Any one of the drop decisions and algorithms is sufficient to be implemented. Consequently, it makes sense to use the method that consumes the least resources in terms of processing power and DRAM bandwidth.

If billing is done on a per-packet basis and the SA is used to assign the current packet count to a policy, then a huge address space is again created. However, most likely billing will be done using a tuple of SA and QoS parameters. With 16 traffic types in the QoS field, this multiplies the table size by 16. If billing is done by a lookup of the policy based on SA, QoS, and policing parameter out of a database, the situation gets more complex; not only is one more lookup added for the billing policy, but size is added to the billing table by multiplying the number of possible combinations by the number of entries derived form the policing table.

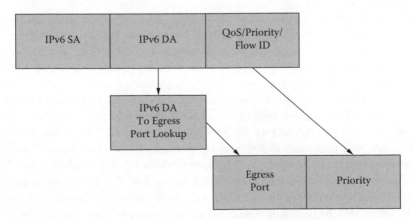

FIGURE 8.7 Algorithmic QoS determination and egress port lookup.

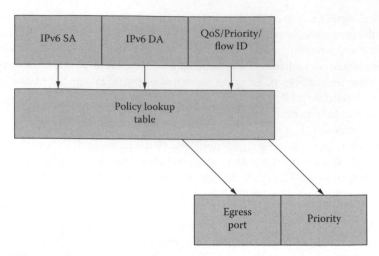

FIGURE 8.8 Database lookup of policy.

As a result, reducing the number of policies and entries in the policing and billing databases is essential. Both tables must be reduced to a reasonable size for the lookups to be done and for the table entries to be written back in a timely fashion. Additionally, it must be ensured that the number of successive lookups is minimized so the end result of the lookup can be delivered in time. It does not make sense to use the LCI to determine a pointer to a pointer to a database entry because it involves unnecessary lookups. Instead, the LCI could be used to directly determine the policy, either algorithmically or by a single direct lookup. Every successive lookup adds to the latency of the availability of the final lookup result, and that latency will contribute to the delay and delay variation of the datagram traversing the router. It will also require deeper buffers and queues to avoid dropping a datagram because its policy could not be assigned in time. A consequence out of these facts is that database lookup performance is crucial to the performance of the network processor and traffic manager.

Policing is always carried out on the basis of the external datagram and for the external datagram, even if the network processor or the traffic manager processes the internal datagram. In this case, the network processor and the traffic manager must perform the same decisions across a set of internal datagrams belonging to the same external datagram.

Policing allows checks on access control lists for each and every incoming packet and therefore allows access control to certain resources based on access policies. The advantages of implementing ACL within a router instead of within the server means that malicious code that would circumvent server-based security will not impact the router and therefore will not compromise the server. Additionally, not only ACLs can be enforced, but also certain ports, such as TCP ports, can be blocked. They can be blocked just because of the port number, but additionally, they can be blocked based on port information together with other pertinent information. So a router can block all TCP ports that pertain to remote management of a range of computers' IP addresses if they are not associated with a particular IP address. Mostly, the line cards support policies that allow the creation of rules—functions of certain parameters that can be logically related and concatenated in a variety of ways. These rules are enforced by the line card and its processor under its operating system. As a result, if it does not execute any malicious code that would compromise the server, then it will make intrusion into the server more unlikely, and therefore the server will be more secure. These rules can apply to TCP ports, UDP ports, a mixture of them, or a subset of any of them. It can apply to those without any further conditions, or with additional conditions. Policing has a lot of different implications. While it will consume computer resources on the line card's processor, it diminishes load on the server's processor. It reduces unwanted traffic in the protected parts of the network because it will drop traffic before it enters the protected part of the network—typically the internal part of the network. It reduces the total traffic, and it significantly reduces malicious traffic.

As we can see, policing can contribute to data loss. For low-priority datagrams that are timing-insensitive, retransmitting them is a viable solution. For TDM-like traffic, this is not a solution, which is why policing is crucial to unifying networks. However, not instituting a drop policy for excess traffic or traffic that exceeds quotas would result in a proportionally higher loss of data—possibly of high-priority data covered by QoS guarantees or SLAs.

It must be stressed one more time that dropping traffic due to violation of policies such as access rights or executing a forbidden operation is different from dropping traffic because of traffic shaping for management. While traffic management at any cost tries to avoid loss of high-priority traffic that is delay-sensitive, dropping traffic for violation of policies is insensitive to the priority of the traffic. As a result, completely different and separate units or software modules process traffic management and traffic policies. There should not and must not be any connection between the two modules.

DATA LOSS AND HIGHER LAYERS

Higher layers will take care of the traffic loss and eventually request a retransmit, and therefore contribute to an even higher total amount of traffic—but at least it will not occur when the rest of the system is in overload anymore. A temporary overload situation can be resolved that way, but a better solution would be to have traffic management queue and buffer the temporarily excess traffic. If traffic has exceeded a user's quota, then it will be permanently blocked until the quota is reset or the limit is increased. In those cases, the retransmit request will not result in successful data transmittal. Permanent or persistent link outages cannot be ignored, and traffic shaping is not designed to take care of these situations. Traffic engineering is required to redirect high-priority and real-time traffic flows around persistently or permanently compromised links. Traffic will have to be rerouted for the duration of the link outage or the duration of the reduced link throughput, and either SNMPv3 or Q3 messages to the NMC can accomplish computing new optimum paths and their setup. In any case, this is a traffic engineering function of the NMC, not a traffic-shaping function that can be handled locally in the traffic manager.

Requirement 4: perform Traffic Management by queuing and buffering traffic according to SLAs, and drop excess traffic. Traffic Management is a very important feature of a router, whether it is deployed in the core of a network or at the edge. In the core of networks, links are the expensive resource. While dark fiber—once laid—is cheap, dim or lit fiber is not. Independent of whether SDH/SONET is the underlying frame structure using Packet over SONET (PoS) or not, IP over dim fiber without any additional framing uses the carrier's infrastructure and therefore must be paid for. As a result, the line with its transmission capabilities is leased and paid for—independent of whether the available bandwidth is used or not. At the edge, typical limitations of available bandwidth force the administrator to make smart use of the bandwidth by shaping traffic. Consequently, a better usage of the available bandwidth makes a significant difference in the cost structure of the network backbone. Better usage is based upon shaping the traffic and thereby managing it. Traffic shaping can occur if excess traffic is of a type that allows temporary queuing. This excess traffic is queued until spare transmission capacity is available again, and then sent out. For this to happen, traffic must be of different types, and at least one of these must be invulnerable against variable and unpredictable delay and delay variations. See Figure 8.9 for an example of unshaped traffic.

The ingress-side network processor derives the Local Connection Identifier (LCI) out of the information in the headers and the payload of the datagram, and in some cases will, through lookups, use internally available information to derive the LCI based on that information, either directly or indirectly. The LCI contains at least the destination port address. Mostly, it also contains aging descriptors, the ingress port number for reassembly, a cell sequence number, and the priority of the datagram or any other descriptor for its requirements regarding real-time forwarding. The traffic manager will use this information to determine if the datagram must be forwarded to the switch fabric immediately. If that is the case, then it takes no further action on this particular datagram. If the datagram can be queued and the egress-side

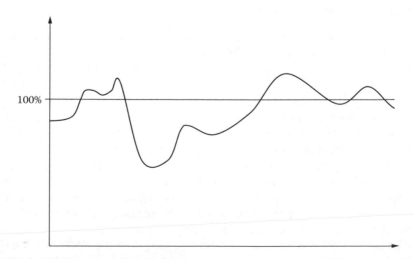

FIGURE 8.9 Unshaped traffic-offered load.

link capacity is exhausted, then the datagram can be enqueued. The link capacity status is available to the traffic manager by means of internal communication between the egress-side network processor and the egress-side traffic manager.

This is of course very undesirable. Just clipping the excess traffic results in a large number of datagrams being discarded (see Figure 8.10), and the datagram drop decision is made based on its arrival rate, and not on its priority, real-time requirements, or any other useful measure. The excess traffic is depicted in Figure 8.11, and it is clear that other drop decisions must be made to not trigger an undue amount of retransmit requests.

The datagram arrival rate at the ingress line card cannot exceed 1. The datagram departure rate at the egress line card cannot exceed 1, either. However, the offered

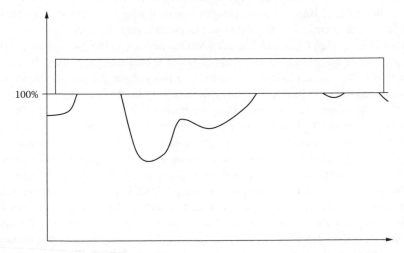

FIGURE 8.10 Shaped (clipped) traffic.

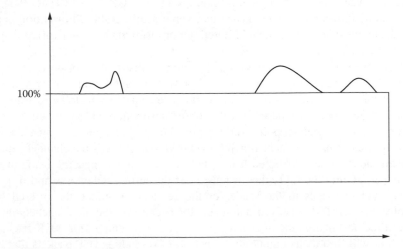

FIGURE 8.11 Excess traffic.

traffic towards the egress card's port may very well exceed 1, and that is where traffic management becomes important. In that case, the traffic manager must enqueue excess traffic according to its priority and real-time requirements, and institute drop policies to reduce persistent excess traffic. Why can the offered load from the switch fabric exceed 1 at the egress port? Because a packet or cell switch typically does not have Call or Connection Admission Control (CAC) capabilities, and therefore each and every packet or cell carries its own header containing at least the destination address. This destination address may serve as the only information the router has about the datagram, and therefore will match the destination address with a port that forwards the datagram to its final destination or the next hop towards it. It is not subject to any kind of bandwidth control on the egress port. If all links from the switch fabric to the line card operate at the rate of the external link, then the datagram arrival rate from the switch fabric to the line card is identical to the datagram departure rate from the line card to the external link. In that case, the line card would never receive more than 1 Erlang of traffic from the switch fabric. This is not a useful design if there is a traffic manager on the line card. In those cases, the links from the switch fabric to the line card and from the line card to the switch fabric should have some "overspeed" over the rate of the external link. This "overspeed" is provided to allow the line card to fill up the traffic manager's queues even while the departure rate of datagrams from the line card to the external link is 1 (or 100% of its link capacity). This is useful to be able to provide non-timing-critical datagrams to the queue so that the traffic manager has enough data in the queue to fill the egress link if the arrival rate of datagrams at the traffic manager falls below the departure rate of datagrams. This ensures that the traffic manager always has enough data available to fill the egress link. It also helps to increase the net data rate into the traffic manager over the egress link's data rate for internal messaging that might take up some capacity on the link between the switch fabric and the traffic manager. Some signaling is inband (within the datastream, with different protocol headers to distinguish user data from signaling data), such as EGP, BGP, CIDR, RIP, and queue status information. This means that some routing table information is sent inband, and must be identified and filtered out or inserted by the network processor on the line card or processor card.

As a result of internal overspeed and stochastic distribution of destination port and priority tuples, any egress port may temporarily be overbooked or even flooded. In the worst case, all ingress ports of a router receive datagrams to be forwarded to the same egress port. A random distribution of destination addresses will result in temporary random egress port overbooking. The traffic manager must enqueue excess traffic during periods in which the offered load exceeds the channel capacity and dequeue it while the offered load is below the channel capacity. If the offered load is persistently too high, then it must start discarding datagrams and trying to send inband messages to the sender or the senders to reduce the offered load. Partially, in TCP this is solved automatically because if the TCP packets do not arrive at the destination, the intended receiver does not send a TCP ACK back, and the sender will throttle its output after a period of resending TCP packets that have not been confirmed received.

Another contributing factor is that the channel capacity of the egress link may very well be *below* 1 for extended periods of time. If the node or the terminal device on the receiving side is in overload, then it will start rejecting traffic until its overload situation is resolved. This overload may be because its own egress link to a certain destination is throughput-reduced, or because of internal congestion or contention. As a result, its ingress ports will delay or deny incoming traffic, thereby reducing its ingress link's transport capacity. Since that is the sender's egress link, the sender must be able to cope with this situation. It must shape the traffic, discarding traffic that persistently exceeds the transport capacity.

Egress line contention is one of the hardest obstacles a line card must overcome. It is unknown to the line card and the traffic manager if the contention is temporary or permanent. If it is temporary, then enqueuing more non-real-time traffic makes sense because it can be dequeued at a later point in time—if the queue capacity is not exceeded. This enqueued data will have to be discarded if the time to live is exceeded, the egress link capacity remains reduced, or if the capacity of the traffic manager's DRAM is overrun. In all these cases, the traffic manager must spend some effort to discard the enqueued data while dequeuing whatever is possible and executing drop policies on newly incoming data.

Traffic management and traffic shaping typically include algorithms to enqueue excess traffic into the traffic managers' DRAM during times when the offered load is higher than the egress channel capacity, and to dequeue it from the traffic managers' DRAM during times at which the offered load is lower than the egress channel capacity, so the channel capacity is fully utilized. Enqueuing and dequeuing both is performed based on the destination port address, the priority, the real-time requirements of the cell, and on information the traffic manager has gathered about the egress link situation. In some cases, the traffic manager additionally holds a dynamic backpressure list on a per-port, per-priority level, so overload can be avoided without having to assert overflow conditions. Both traffic management and traffic shaping try to use a queue memory to buffer traffic that exceeds the egress link capacity. However, due to conditions on the egress link that are not under control of the traffic manager, the situation may be such that entries age in the queue memory age and are outdated by the time they can be dequeued, and therefore must be discarded. This may happen at the time when the datagram is dequeued, but it is smarter to perform the datagram drop decision at the time it is enqueued. Whenever it is foreseeable that a datagram may not be forwarded right away or dequeued by the time required in its LCI or other tag, the traffic manager should discard the datagram. Such an algorithm avoids using processing power and DRAM bandwidth for datagrams that ultimately must be discarded anyway. However, it is not always possible to do so, and therefore a discard algorithm is required on both the enqueuing and dequeuing side of the traffic manager. It is important to note there are two reasons to discard traffic. One is persistent egress link or internal congestion, contention, or overload that cannot be overcome by queuing. The other reason to discard traffic is exceeding quotas. Both can and should be performed by the traffic manager. The drop policies will be looked up by the network processor and indicated in the LCI. The overload situation is signaled by backpressure from the switch fabric or by inband messages from the egress-side traffic manager.

In most cases, traffic shaping for traffic management purposes is performed into the egress link. In rare cases, it will occur in conjunction with ingress port side traffic shaping in order to load the switch fabric as much as possible, and to achieve link load parameters of about 1 Erlang. However, it must be clear that the traffic manager in any case cannot and will not replace the switch fabric's queue manager. This confusion can come up especially with ingress-side traffic management and shaping. Traffic management is intended to shape traffic by enqueuing, dequeuing, and executing drop decisions. The drop decisions must be made as soon as possible in order to not use up resources such as processing power and DRAM bandwidth. In some instances, not only must the egress link be loaded to the highest possible extent, but also the switch fabric of the router. In those cases, it makes sense to shape also into the switch fabric. This ensures that the switch fabric is loaded as much as possible. The traffic manager on ingress cannot displace the queue manager of the switch fabric. The queue manager of the switch fabric has a different scheduler, and therefore provides significantly lower latency into the crossbar switch of the switch fabric chipset. While the traffic manager has a much larger capacity of datagrams to be queued, its latency is higher, and therefore it cannot make sure that the crossbar switch is loaded to the maximum possible extent on all channels. The combination of ingress-side traffic management and a queue manager on the switch fabric card provides the highest throughput and load, and therefore the highest utilization of the switch fabric—and as a result, the lowest probability of stalling the input side of the egress-side traffic manager.

DISTRIBUTION OF TRAFFIC

If an equal distribution of packets lengths is (wrongly) assumed, then there will be an equal number of packets for each length over time. However, that is not how the distribution of packet lengths works. In reality, the sender's line card or NIC will restrict the packet lengths to more manageable sizes. One of the many restrictions is the cluster size of files on hard disks. This is limited for most operating systems to 512 bytes. FAT16, FAT32, and NTFS file systems support 512 bytes per cluster. As a result, we will have an equal distribution of packet lengths from 1 to 511 bytes, and a large peak at 512 bytes. The height of the peak only depends on the maximum file size allowed by the operating system, since it is determined by the integer part of the dividend of the largest file sizes and the cluster size. A source file length maximum of 1.5 kByte leads to a peak of the 512-byte packet length—three times larger than the rest of the equal distribution. As a result, we must optimize the router to an equal distribution of packet sizes from 1 byte to the size of the cluster minus 1 and a peak at the cluster size. The more pronounced the peak, the better the efficiency of a cell-based line card and a switch fabric. In other words, if the peak is at 512 bytes, then a cell-based line card and switch fabric is optimized towards 64-byte sized payload cells. Obviously, the 512-byte peak can be handled perfectly in a line card or a switch fabric with 64-byte payload cells. The equally distributed number of fragments ranging from 1 byte to 511 bytes again falls into equal distributions of fragment sizes from 1 byte to 63 bytes, and a peak of 64 bytes that is n times as high as the rest. For all 512-byte fragments, the efficiency of the cell-based line card and switch fabric is 1.0, and for all fragments in the range of 1 byte to 63 bytes the efficiency converges around 0.5. The only impact on the global efficiency of the cell-based line card and switch fabric is the relative height of peak versus the rest of the equally distributed fragments. This becomes very apparent in Figure 8.12.

In the example of Figure 8.12 the number of cell-sized fragments N_C and the number of fragments N_F in size between 1 and the internal cell size minus 1 both are a function of the internal cell size L_C and the largest allowable packet size L_P.

From this equation is derived that the internal cell size significantly impacts the efficiency of the line and the switch fabric. The distribution of the internal size of the internal cells to be forwarded through the switch fabric depends on the fragment length L_F ranging from 1 to the internal cell size minus 1 and the number of the fragments having the internal cell size.

If $L_P = L_F + n * L_C$, with $L_F = 0$ or $L_P \bmod L_C = 0$, then the number of fragments N_F sized from 0 to the internal cell size minus 1 is the modulo of the largest allowable packet size L_P and the internal cell size L_C each, or $N_F = L_P \operatorname{div} L_C$, and the number of fragments having the cell size $N_{F,C}$ is given by $N_{F,C} = \tfrac{1}{2} * L_P \operatorname{div} L_C * (L_P \operatorname{div} L_C + 1)$. If $L_P = L_F + n * L_C$, with $L_F \neq 0$ or $L_P \bmod L_C \neq 0$, the number of fragments N_F sized from 0 to $L_P \bmod L_C$ is the modulo of the largest allowable packet size L_P and the internal cell size L_C each, or $N_F = L_P \operatorname{div} L_C + 1$, the number of fragments N_F sized from $L_P \bmod L_C + 1$ to the internal cell size minus 1 is the modulo of the largest allowable packet size L_P and the internal cell size L_C each, or $N_F = L_P \operatorname{div} L_C$, and the number of fragments having the cell size $N_{F,C}$ is given by $N_{F,C} = \tfrac{1}{2} * L_P \operatorname{div} L_C * (L_P \operatorname{div} L_C + 1)$.

In any case, for all packet lengths greater than twice the cell size, the number of cell-sized fragments is larger than the number of fragments for any other size.

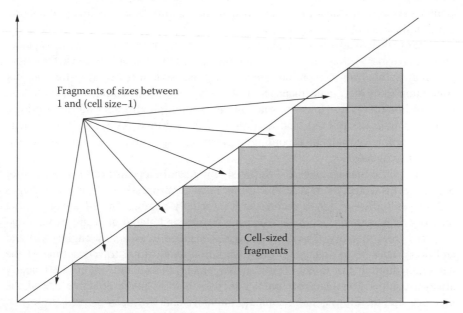

FIGURE 8.12 Fragments versus packet lengths.

As a result, in this case the efficiency of a cell switch is greater than 0.5 under all conditions.

The realistic case is that the two maxima are at 64 bytes for all the HTTP put commands and 512 bytes packet length for any file transfers, and taper off very quickly after that. As a result, the efficiency is skewed even more towards 1 because of a further predominance of the 64-byte fragments and cells. We can therefore assume that the efficiency of a cell-based line card and switch fabric is greater than 0.5 and less than 1.0—but it is closer to 1.0 than it is to 0.5.

This is a huge improvement over packet-based line cards and switch fabrics, where the efficiency becomes smaller when the range of the distribution of the packet of fragment sizes gets larger. The efficiency of these packet-based line cards and crossbar switches is significantly below 0.5, if not below 0.10. Longer packets therefore seldom affect the average traversal time of packets through a switch.

TRAFFIC MANAGER FUNCTIONS VERSUS QUEUE MANAGER FUNCTIONS

Many router designers misunderstand the purpose of the traffic manager on the line card or processor card and the purpose of the queue manager of the switch fabric chipset on the switch fabric card. They fulfill fundamentally different functions.

The traffic manager typically shapes into the egress link or channel. Its channel capacity is essentially constant. Therefore, its utilization largely depends on the offered load. To a much smaller degree, it depends on the ability of the receiving side to accept traffic. As a result, the utilization of the channel is at its maximum if the traffic manager always has some enqueued data that is not timing-critical and is of low priority in its buffer—ready to be dequeued and sent whenever there is not enough offered load from the switch fabric to fill the transport capacity of the channel. As a result, the traffic manager that shapes into the egress link will achieve a link utilization of about 1, or 100%.

The queue manager has a fundamentally different purpose. The channel capacity into the crossbar switch part of the switch fabric is highly variable because the egress port may suffer from contention. Since a crossbar switch is blocking-free but not contention-free, the queue manager must only handle preventing contention and Lead-of-line blocking. However, since there are multiple different traffic priorities, it must handle the offered load according to its priority. As a result, the queue manager will have to queue according to priorities, and provide per-egress-port and per-priority queues.

Both traffic management and traffic shaping can be performed on a per-cell or per-packet basis. In a lot of cases, the traffic manager processes cells, i.e., it manages and shapes traffic on egress before the Reassembly engine. The advantage is that memory management is easier because fixed-sized datagrams are stored and retrieved from memory. The disadvantage is that the Reassembly engine will add additional latency and datagram delay variation to the total traversal time of the datagram through the router. Traffic management and shaping of packets nearly always requires very elaborate pointer management to make effective use of the memory, and therefore is more compute-intensive and will need garbage collection routines running periodically. The advantage is that the traffic-shaping process does not impact the Reassembly times anymore because Reassembly has already occurred before within the traffic management engine.

Ideally, the egress-side traffic manager performs Reassembly. This combines the efficiency of storage of cells coming into the traffic manager with the ability to shape traffic and discard packets (or more generally, egress datagrams) more efficiently as an entity and not on a per-cell basis. Combining SAR with traffic management on egress saves resources and processing effort because the Reassembly engine can reuse memory from the traffic management engine, and at the same time is in better control of the datagram delay and its variation. Additionally, the datagram delay variation caused by Reassembly can be controlled better than if those engines were separate and Reassembly occurred after traffic management is performed. Combining SAR and traffic management on egress reduces the adverse impact of the Reassembly process and reduces its impact on the delay and delay variation of the traversal of datagrams through the line card, and therefore the router.

Also, shaping on a basis of the external datagram instead of the artificial internal datagram will render better results because of better efficiency—the shaper has insight into the external datagram, and not only into the internally used format. As a result, shaping the reassembled external datagram makes more sense.

INGRESS-SIDE TRAFFIC MANAGEMENT

The ingress-side traffic manager does not have access to the information about the load status of the switch fabric other than the backpressure bitmap, nor has it any information about the queue status of the egress-side traffic management. It will require this information to determine what to do next with the datagram. If there is no switch fabric overload indicated by the lack of backpressure signals and the traffic manager does not have to drain higher priority datagrams than the ingress datagrams, then these can be forwarded to the switch fabric. If the switch fabric, the traffic manager, or the egress-side link is in overload and the datagrams can be queued, then the traffic manager will have to enqueue them. If those datagrams are timing-sensitive, then the ingress-side traffic manager will have to discard them. This depends on internal statuses of the router—more than it is the case for egress-side traffic management. While it makes some sense to have ingress-side traffic managers, egress-side traffic managers are more useful. Ingress-side traffic managers load the switch fabric to the maximum possible extent, but cannot guarantee maximizing of egress link utilization. Egress-side traffic managers can make sure that the egress link utilization is maximized for as long as there is enough offered traffic. If there is enough space on the line cards, and if the power budget allows for it, then it makes sense to use ingress-side and egress-side traffic managers. In any case, both the ingress-side and egress-side traffic managers will have the information they need for processing the datagrams, namely their destination port address and the priority level, from the LCI which was determined by the ingress-side network processor. Using traffic management on ingress and on egress also requires internal communication between the two traffic managers about the queue status.

The non-shaping line card (see Figure 8.13) simply does not include a traffic manager—neither on ingress nor on egress. The traffic-shaping line card (see

Figure 8.14) contains an egress-side traffic manager, and it may contain an ingress-side traffic manager.

Connection Admission Control (CAC) is useful in routers that must set up connections—even if these are only logical connections or virtual connections (see Figure 8.15). In any case, they contribute to the ability of the router's line card to reserve bandwidth and avoid contention within the switch fabric and the egress link for the amount of data that is covered by the CAC contract. MPLS, RSVP, any TDM-like traffic that relies on bandwidth reservation, tunneling any traffic with bandwidth guarantees, and fixed-bandwidth video distribution will require CAC or similar mechanisms. It is unnecessary in routers that do not have to support any of the above (see Figure 8.16). CAC has a fundamental problem associated with it. Since it reserves bandwidth for a logical or a physical connection, this bandwidth cannot be reallocated when it is not utilized. As a result, CAC is not an efficient way to unify networks that carry data and TDM-like traffic.

Requirement 5: support Traffic Engineering by means of collecting Statistical Traffic Data. Traffic Engineering is used to determine traffic flows and optimize the infrastructure such that it can switch, route and forward the offered load at the minimum cost to the carrier or ISP. In order to do so, traffic flows must be recorded and ultimately processed. This requires each router in the network to collect statistic traffic data and forward this information to the Network Management Center (NMC). The router collects this information in all line cards, processor cards, port cards, and every other component, subsystem, or module. This collected information is then sent to the OAM&P card, consolidated within the router or node, and sent to the NMC. The information collected will have to contain traffic types and priorities, whether the datagrams were queued or not, and whether, where, when, or why they were dropped. The cards will have to record the egress link status along with

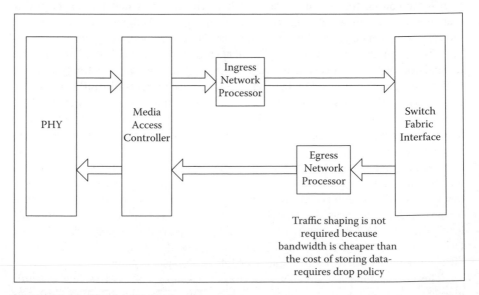

FIGURE 8.13 Simple non-shaping line card.

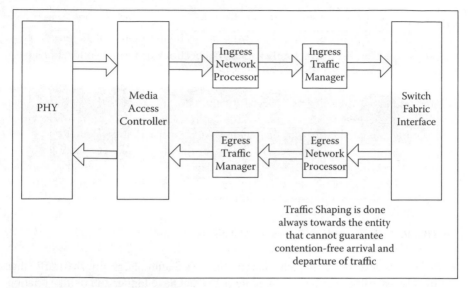

FIGURE 8.14 Traffic-shaping line card.

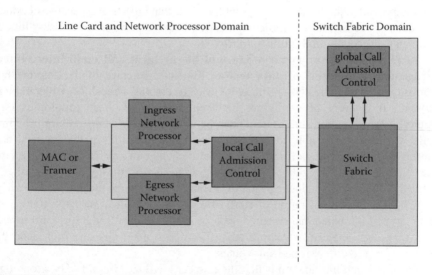

FIGURE 8.15 Router with Connection Admission Control.

the status of the internal congestion or contention in the switch fabric. Depending on the requirements of the NMC, the nodes will have to collect even more statistic traffic data, plus their drop probabilities from policy databases, and other statistics that may pertain to the queuing and dropping. The OAM&P entity might—based on NMC instructions—collect all statistics on all network layers, and even on intermediate layers, if necessary and available. If the NMC has the information about all the routes that datagrams take, then it can make sure delay-sensitive high-priority

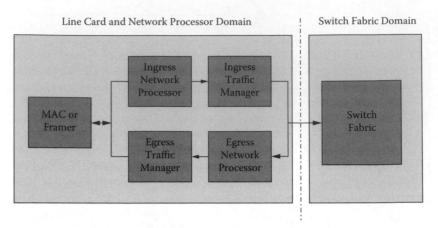

FIGURE 8.16 Router without Connection Admission Control.

traffic always gets routed on paths that provide this feature. High-priority traffic that is not delay-sensitive can be routed along paths that have higher cell delay variation but a small number of hops. Low-priority traffic that is delay-insensitive can be routed along the lowest-cost paths that may have some probability of datagram loss. More importantly, the NMC can pinpoint locations that drop traffic because of policy violations or excess traffic. Traffic engineering helps to prevent unnecessary loss of datagrams, especially high-priority traffic.

Requirement 6: support a mixture of hierarchical and mesh interconnect infrastructure in terms of data traffic. The infrastructure of the Internet has changed quite dramatically, and it in no way resembles a mesh or a hierarchical network. It is more of a mixture between the two of these topologies (see Figure 8.17), and as a result, routers must be able to handle this structure. The line cards will have to be able to determine its communications paths for next-hop routing, and for rerouting of traffic in case of an outage of any links connecting the line card to any of its neighbors, whether hierarchically or through the partial mesh. While this appears to not pose a problem, rerouting will change the datagram latency, throughput, and administration of traffic flows. It also will impact the routing of traffic around the failed and rerouted node or link. As a result, the line card should have a topology map in its routing table in order to determine the impact on the datagrams of the different paths or routes.

The important implication is that the line card will require not only a topology map of all other nodes in its neighborhood, but also a map describing the interconnects between the nodes in order to be able to predict the impact on the number of hops, delays, and delay variation parameters for all routes that are currently in use by any datagrams that traverse the node or one of its neighbors. Rerouting any of those paths must be possible for all datagrams traversing the router with the least impact on all SLA and QoS parameters. This is only possible when and if the router's line card has the ability to draw conclusions from the topology map of its neighboring nodes; it will be doing so in conjunction with the NMC and OAM&P entity.

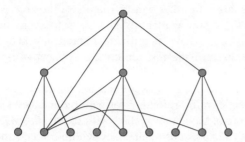

9 terminal nodes,
3 intermediate nodes,
1 top node. Result is
4/13 overhead.
Direct connections
between heavy traffic
nodes.

FIGURE 8.17 Superimposed mesh and hierarchical network.

Requirement 7: support an overlay network for metadata and signaling data. Nearly all signaling data is embedded in the data stream, and it must be detected by the network processor by its type and embodiment. As is true for all other data, this metadata must be parsed and processed accordingly. If it is terminated in the parsing line card, then it must be removed from the data stream and processed. In some cases, an incoming command will trigger sending a message back, or sending a query or a command out. If metadata is sent out, the originating line card will have to insert this metadata into the data stream to the recipient. This metadata may also contain instructions on how to treat the immediately following datagrams, datagrams with the same or similar tag, or all subsequent traffic. It can be an instruction for multicast, broadcast, tunnel setup, tunnel teardown, or any other filtering command for tags, headers, or datagram contents. It can be routing data (routing table entries or updates thereof), and it can be any other kind of setup instruction that is not sent and distributed through OAM&P commands. This overlay network is a virtual overlay network, and the signaling metadata may be associated with other destination addresses, other forwarding tables, or other Access Control Lists. Metadata and signaling can contain EGP, BGP, RIP, and CIDR messages, but also more abstract messages for load balancing or equivalent messages. These messages can be transmitted through inband messages to the line cards or through explicit messages to the OAM&P entity. These methods are not mutually exclusive and may complement each other. In any case, these messages terminate in the line cards; independent of whether they were received by the line card directly as inband messages or through the OAM&P entity. The line cards can also be the sources of these messages, and they will use the same mechanisms to send those messages out to other routers. They will process these messages and take appropriate action. Metadata and signaling is thereby disassociated from the data impacted. It can be routed on different paths with different priorities using different methods of distribution. The line cards will have to support the sending, receiving, processing and forwarding of metadata and signaling information.

Requirement 8: enable traffic rerouting at the edge and provide redundant fail-safe systems towards the core. While current networks typically rely on path redundancy, this will be impossible with a network that has to withstand DDoS attacks and safely and securely route emergency (E911) traffic. As a result, future networks will have to primarily rely upon device redundancy and secondarily upon path redundancy in the core, and primarily upon path redundancy and secondarily upon device redundancy at the edge. In other words, routers in the core will have to support at least a "five nines" system availability and fail-safe operation, while routers closer to the edge might not have to provide redundancy and fail-safe operation. Routers at the edge that provide mission-critical connectivity—such as routers that connect E911 call centers to the core network—will have to conform to high-availability requirements. Both the software and the hardware on the line cards will have to support dynamic traffic rerouting for path redundancy as well as for device redundancy. In the core of the network, the NMC and the OAM&P entity will therefore have to first try to make use of redundancy within routers, and if that fails, use a different path to circumvent the failed router. In the perimeter, the situation is the opposite. The OAM&P entity and the NMC will first try to make use of path redundancy by rerouting traffic and making the line cards aware of this. These requirements will impact the software on the line cards and the OAM&P entity. Depending on where the router is within the network, and what its immediate logical and topological neighborhood setup, the router's line card and the OAM&P entity will have to either reroute traffic through other paths, or use the internal subsystem redundancy to circumvent failed components and subsystems. In any case, the line card will require the ability to support local rerouting and path swapping. This information is provided by the NMC to the OAM&P entity of the router, and subsequently to the line cards.

Requirement 9: communicate with the PSTN infrastructure. Some line cards will have to communicate with the Public Switched Telephony Network (PSTN). In some instances this will be to connect a router through the PSTN to an ISP, a carrier, or a corporate network. In other cases, it will be to connect the PSTN infrastructure to the data network backbone. In those cases, the router's line card must understand the signaling within the PSTN, and from the PSTN to the end user. The PSTN TDM network is in existence today and cannot just be ignored. This will be especially true once VoIP becomes more common. If the VoIP network wants to make use of the PSTN, then the VoIP gateway within the line card will have to communicate with the PSTN signaling gateways through a method or a protocol such as Common Channel Signaling Number 7 (CCS#7). In general, the line card must understand CCS#7 (US: SS7, Signaling System 7) signaling to communicate with the Class 4 or 5 Central Office switches of the PSTN. It also must be able to understand ISDN, B-ISDN, the ISUP, the IN infrastructure, its signals and messages, as well as its statuses. For 800 number routing and 911 routing, it must be able to access the features of the Intelligent Network (IN) and Advanced Intelligent Network (AIN). The line card also must be able to handle local number portability (LNP), number portability and the PSTN-to-Cell phone internetworking. It will additionally have to support all user signaling like caller ID, CLIP and any other signaling that is in use today. For CCS#7, understanding the messages and commands provided by the metasignaling is essential. It will also require the ability to generate or forward the messages, commands, and the status data of CCS#7, a virtual overlay network for signaling and metadata. The line card in which the metasignaling traffic originates or terminates may or may not be the line card that sends out or receives this data on its associated physical or logical port(s).

FURTHER IMPACT OF ADVANCED ROUTER ARCHITECTURES

While one might think that advanced router architectures impact those nodes and maybe our expectations of the future Internet, the implications go much further. They impact local networks and servers as much as they directly impact the core and the core nodes of the Internet.

LINE CARDS AS SERVER NICS

While we will see different routers and different line cards in the new generations of routers, the impact of changed line card architectures will be beyond just routers. In the future, we will see more and more server NICs that will look like router line cards. The reason for this is the fact that all the data is stored on the servers with the policy information for any particular user or connection. In addition, in the routed networks of the future, access to policies, to routing tables and updates, to DNS, and possibly to other lookup functions will work like today's AIN; there will be more and faster centralized servers providing these fields of information. Routers and servers will be forced to work in a more integrated, comprehensive network view. Another reason to include IPv6 router function in server NICs is that it will become unacceptable to consider 40% a heavy load status on any given link. In the future, a load level close to 100% or 1 Erlang will be considered an acceptable load level, with a 10% high-priority traffic level.

SCALABILITY

One of the main concerns of network architects is scalability. Scalability is defined as the ability of a system to be aggregated with similar other systems to provide higher throughput. Scalability can be linear or non-linear. Linear scalability implies that n systems provide the n-fold throughput of one system. Most systems do not scale linearly, or only to a certain threshold. In a network, scalability can pertain to the network itself, or to each of the nodes. For example, the Internet scaled (non-linearly) from a few users to several million users by increasing the number of nodes and terminal devices. One could argue that it was scaled closer to a linear increase if the nodes had grown in throughput instead of increasing in number. This is happening today. The number of nodes is in fact being reduced, but the nodes grow significantly in their throughput, making the Internet more manageable, more stable, and involving fewer hops on average to route packets from their source to their destination.

With the new generation of advanced routers, the Internet will maintain its scalability into the future. The new generation of advanced routers provides a proportionally higher throughput than older generations, and therefore reduces the number of interconnects, the proportional amount of required signaling traffic, the number of hops, and the depth of hierarchy compared to the PSTN. While most of the scalability is due to the switch fabric, a large portion of linear scalability in a node is due to the port throughput and the number of ports on the line cards. Functions and features of the line card, such as linespeed lookups, policing, and traffic shaping allow for growth of the line card throughput.

DIFFERENTIATION OF FUNCTIONS

As we can see, different subsystems of the router and even within the line card use different portions of the LCI (see Figure 8.18). None of these subsystems or components must use any other information. This information is extracted by the network processors and prepended to the payload as a tag to guarantee all subsequent units and engines have the same information available to perform their functions, and that no contradictory information is used for switching, routing, forwarding, queuing, or discarding the datagrams. While the switch fabric only requires the destination port number and the priority, the traffic manager must know the destination port number, the priority, the real-time requirements, the drop policy, and the source port number. The traffic manager will enqueue and dequeue based on the egress port number and the priority within the LCI, based on additional router-internal information such as its queue status and the egress link status. However, it does not use any other information from the datagram.

The same is true for the policing engine. It receives the drop policy from the network processor and determines when and if to drop the datagram. The drop policy can be based on authorization for the use of resources, exceeding thresholds of usage, temporal requirements, group associations, or any other logical operation. The policing engine will retrieve the necessary information from the LCI and the policy database through the network processor. The Reassembly engine needs the source port number, the priority, and the Cell Sequence Number (CSN)—but no other information within the datagram—to correctly reassemble a datagram and make a decision about when to abort a reassembly attempt based on elapsed time from the arrival of the first cell of the datagram.

All engines can be implemented in hardware modules or as separate module building blocks in software. In any case, they must perform their functions separately and independent of each other, and in a logical and reasonable sequence. It does not make sense to determine whether to enqueue a datagram if the policing engine has

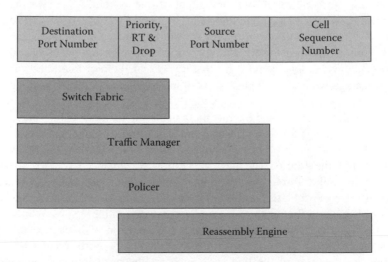

FIGURE 8.18 LCI and its components.

decided to drop it because it belongs to a senders' IP address that is permanently blocked by a policy.

LINE CARD IMPLEMENTATION IN SINGLE-BOARD DESIGNS

In single-board designs, all functions are on the mainboard, including the line card logic. Typically, there is one logical and physical line card equivalent—consisting of the PHY, the MAC, the optional framer, and the network processor—for each physical port of the router (see Figure 8.19). Because of restrictions of PCB area on the PCB, functions that consume a lot of PCB area are not implemented. As a result, single-board designs practically never contain traffic management functions. Very often, CAMs to speed up lookup processes are not implemented either. Instead, these functions are implemented in software on the CPU or network processor of the line card. Mostly, even in single-board designs, there is one CPU or network processor per physical port because of the massive amount of processing required.

LINE CARD IMPLEMENTATION IN MIDPLANE DESIGNS

Midplane designs differ significantly from the single-board designs. First of all, they provide a lot more space for all the functions that must be implemented. Second, they partition the functions such that they typically provide a port or interface card and a processor card. Strictly speaking, there is no physical "line card" for midplane designs. In midplane designs, the line card is a logical entity, consisting of the port or interface card and a processor card (see Figures 8.20 and 8.21). Each of these cards is larger than the space available for the port or interface and processor functions on single-board designs. As a result, more functionality can be

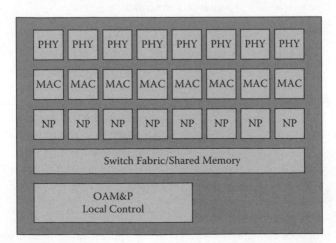

FIGURE 8.19 Single-Board router's mainboard with line card equivalents.

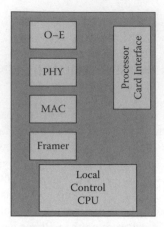

FIGURE 8.20 Port or interface card of a midplane architecture type router.

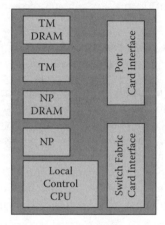

FIGURE 8.21 Processor card of a midplane architecture type router.

implemented. In midplane designs, all of the Line Specific Logic is located on the port or interface card, and all the processing is performed on the processor card. A consequence of this architecture is the fact that the processor cards are common to all port or interface cards, and therefore one single design of the processor card hardware can be used for all different port cards. This makes the design of the processor card hardware cheaper, since its hardware design cost can be shared among the wide variety of port cards. However, the software design for the processor cards is unique to each of the port cards supported by the processor card.

LINE CARD IMPLEMENTATION IN BACKPLANE DESIGNS

Even more pronounced is the difference in backplane designs. In these designs, the port or interface and the processor cards are physically located on the same

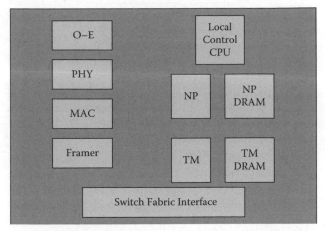

FIGURE 8.22 Line card of a backplane architecture type router.

Printed Circuit Board (PCB). This means they offer the same amount of surface area as the port or interface card and the processor card together in midplane designs. However, this layout offers more flexibility in terms of placement and usage of the area. More importantly, the processor card hardware can more easily be adjusted to the specific need of the processing task. Therefore, their hardware and software can be tailored to specific application needs. As a result, their hardware and software is significantly more optimized towards the functions they must perform, and less general-purpose.

Additionally, with the larger PCB area available, it is possible to implement multi-port line cards in backplane designs (see Figure 8.22). This reduces the per-port cost and allows for a larger number of ports in a router, and also enables local switching, and therefore a higher concentration factor. Multi-port line cards may contain one network processor, one traffic manager, and one lookup engine for all ports on the card. They may alternatively contain multiples of each if required. However, the multi-port line cards contain multiple physical ports or interfaces, and therefore multiple MACs and multiple PHYs. Whenever framing is required, they may contain one or more framers. In some cases, oversubscription is desired. In those cases, the aggregated bandwidth of the ports on the line card is higher than the bandwidth from the line card to the switch fabric card (see Figure 8.23). The assumption is that some of the traffic is local and therefore does not have to be forwarded to any other line card, nor consequently the switch fabric card. It requires that the line card can make local routing and forwarding decisions.

While this is more complex than a standard line card, it has its advantages. It is, of course, not strictly blocking-free because of the oversubscription, but typically that can be handled within the traffic manager. It effectively acts as a router on a line card with a high capacity uplink to the switch fabric card. This reduces the per-port cost and the cost per routed bit, while maintaining all QoS and SLA parameters.

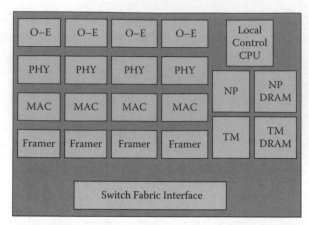

FIGURE 8.23 Multi-port line card with aggregation.

LINE CARD MESSAGING AND COMMUNICATIONS

Line cards communicate with external and internal entities. They communicate with the switch fabric, with each other through the switch fabric, and with external resources through the interface or port card or the functional equivalent thereof on the line card. They also communicate with the OAM&P card, either through dedicated links or inband through the switch fabric. The information sent and received by the line cards is payload data, signaling data, metadata, and OAM&P data. The reasons that line cards communicate are different in nature and priority. As a result, the designer must implement appropriate means of communication between line cards and other entities.

Line cards send all internal datagrams together with an LCI to direct the switch fabric regarding where and with which priority to forward datagrams. Line cards communicate with the switch fabric in order to be able to reduce congestion and contention. For the most part, the switch fabric sends backpressure bitmaps back to the ingress-side line cards to signal temporary overload on a destination port and priority tuple basis. Line cards communicate with each other through the switch fabric for a variety of reasons. They do so inband, i.e., they send messages through the switch fabric to the receiving line card, with a special tag so that the receiving line card can identify the tag and filter out these messages for internal processing. This occurs partially to exchange routing table information, link status, and queue status information in the traffic managers. The communication with the switch fabric is mostly limited to backpressure information sent by the switch fabric queue managers back to the line cards' network processor or traffic manager.

Communication with external resources will originate, terminate, or be forwarded in the line card or the processor card, and will be sent and received through the interface or port card or the interface or port part of the line card.

Finally, the line card will communicate with the OAM&P card. It will send statistic traffic data and billing information to the OAM&P card, and it will receive operational status information from the OAM&P card. This information should be sent through dedicated links and not inband through the switch fabric card. Communication with the OAM&P entity can use data path functions or control path functions within the data path. However, if the data path is down or the control path within the data path is down, then OAM&P becomes impossible. As a result, OAM&P communication within the router should be carried out through dedicated communications channels. It has been proven very cost-effective and more than sufficiently reliable to use Fast Ethernet as a means of communication between the OAM&P entity and the line cards, the switch fabric cards, and in between potentially redundant OAM&P entities.

INTERNAL LINE CARD TO LINE CARD COMMUNICATION

All line cards are connected to each other through the switch fabric card. Therefore, all communication between line cards is forwarded through the switch fabric card. This is true for net payload data, any signaling data, status data, messages, and commands. In order to account for this internal signaling and other metadata, the links between the line cards and the switch fabric card provide some additional bandwidth compared to the egress links of the line cards.

Signaling and the exchange of other metadata between line cards is important for the router to avoid contention, congestion, and blocking, and it is required to distribute internal messages that are not distributed by the OAM&P entity. Routing table updates, EGP, BGP, RIP, CIDR, and other network or transport layer protocols require the line cards to communicate with each other. This communication can take up a significant portion of the bandwidth. Some estimates range from 3–10% of all traffic for this type of signaling and message exchange. As a result, the internal links between the line cards and the switch fabric cards must support this additional traffic.

This internal signaling becomes even more important once network traffic is unified and routers carry traffic that is timing-sensitive and timing-insensitive. In those cases, the line card's traffic managers will have to exchange status information regarding their queue statuses. The purpose of this is to avoid flooding traffic managers and egress links with low-priority (timing-insensitive) traffic when there is enough timing-sensitive traffic offered to nearly fill the egress link. This avoids the necessity of dropping high-priority timing-sensitive datagrams and allows for either queuing or early discard of low-priority timing-insensitive traffic. While situations like this are statistical and therefore temporary, overload can occur on the egress-side link or traffic manager. Since a loss of high-priority and timing-sensitive data is not acceptable, the ingress-side network processor or traffic manager must be able to throttle the source or randomly discard low-priority traffic. However, the ingress-side network processor or traffic manager cannot detect egress-side congestion or contention. The solution therefore must come from the network processor on both the ingress and the egress sides. A possible algorithm will include an inband signaling between the ingress-side and the egress-side network processor, signaling a potential overload situation. These messages can be sent inband in the downstream direction, so that the ingress port side network processor can start applying throttling methods to reduce the ingress traffic (offered load to the line card) or start applying discard mechanisms such as Leaky Bucket, Random Early Discard (RED) or

Weighted Random Early Discard (WRED). The preference is to discard only if the offered load persistently exceeds the link capacity.

INTERNAL LINE CARD TO SWITCH FABRIC COMMUNICATION

To exchange queue status information, line cards require a certain amount of communication with the switch fabric. While the line cards typically would not send any commands or status data to the switch fabric, the switch fabric typically sends back information to the line card about its queues and buffers. It usually sends back a backpressure bitmap to the line card on a per-datagram basis so that the line card is informed about the fill status of the switch fabric queues. The backpressure information gives the traffic manager or the network processor on the line card an overview about statuses of the queues, especially in virtually output queued (VOQed) switch fabrics. This information signals backpressure on a per-VOQ basis, and that equals a per-egress port, per-priority basis. It supports the traffic manager or the network processor on the ingress-side line card to determine the status of the switch fabric and its level of contention. The switch fabric backpressure will react faster and fluctuate more than the backpressure information from the egress-side traffic managers or the network processors, but since the switch fabric backpressure is much more of a transient situation, the ingress-side traffic manager or the network processor can help shape the traffic into the switch fabric by adjusting the offered load into the switch fabric's queue manager.

The switch fabric does not terminate any data sent by line cards. As a result, the switch fabric is transparent for data sent by the line cards. It does not evaluate or filter out any messages or commands sent by the line cards to other entities.

Many switch fabrics support an internal 64-byte payload cell, along with an LCI of 8 to 12 bytes. The line cards must support the switch fabric internal datagram format and derive the LCI from data contained in the external datagrams. This will result in a certain link efficiency. With the numbers mentioned above, the payload efficiency is either $64/72 = 89\%$ or $64/76 = 84\%$. This means that about 11% or 16% is lost due to inefficiencies of the internal datagram format. This loss is easily tolerable, especially considering that typically there is a factor of two for overspeed on the links between the line cards and the switch fabric. The remaining net overspeed therefore is 84% of 2, which is a remaining factor of 1.68, or a net of 68% overspeed. This already takes into consideration that the inefficiencies of the SAR process contribute to a 50% loss of efficiency.

LINE CARD TO SWITCH FABRIC QUEUE MANAGER COMMUNICATION

In order to prevent congestion, contention, and Head of Line (HOL) Blocking, the interconnecting entity—mostly a blocking-free and lossless switch fabric—will have to provide internal links accomodating higher bandwidth between the switch fabric and the line card. The datagram arrival rate at the ingress switch fabric port cannot significantly exceed 1, even with an ingress-side traffic manager. The datagram departure rate at the egress switch fabric port cannot significantly exceed 1 either. However, the offered traffic towards the egress switch fabric's port may dramatically exceed 1. The switch fabric can be seen as a multiplexer with an internal speedup that is significantly larger than 1. At any given point in time, one or more switch fabric ingress ports may offer traffic towards any switch fabric egress port. Each switch fabric egress port can be the destination of two or more internal datagrams arriving simultaneously at switch fabric ingress ports, whereas other switch fabric egress ports may not experience any traffic at all. As a result, the offered load towards any egress port on the switch fabric may exceed 1. Since the switch fabric and its internal interconnects can carry between 1 and 2 Erlang, the datagram arrival rate, and therefore the offered load from the switch fabric egress port towards the line card, will from time to time significantly exceed 1 Erlang.

There will be situations in which this overspeed or bandwidth overprovisioning will turn against the system and allow traffic to be switched from more than one ingress port to a limited number of egress ports, thereby exceeding the link capacity or datagram departure rate of these ports. Over a very short period of time, data can accumulate in the egress port-side traffic managers, and they cannot empty their buffers and queues at the rate at which they are filled. This is independent of QoS considerations, and even with priorities applied to traffic, this will cause a problem over time. Traffic will accumulate in the traffic manager queues and buffers, and it will do so at a rate which cannot be handled any more. If the datagram arrival rate persistently exceeds the departure rate, the traffic manager will have to start applying backpressure and start dropping low-priority datagrams.

LINE CARD TO EXTERNAL RESOURCES COMMUNICATION

Line cards communicate with external resources. Mainly, they transmit net payload data. A portion of the traffic is signaling traffic. This signaling traffic is required to keep the network functioning. It consists of messages, commands, and status data for internetworking between nodes.

The line cards also will have to set up tunnels for IPv4 through an IPv6 network and vice versa. IPv4 routers' line cards will have to set up MPLS tunnels, recognize RSVP, and act accordingly. They are also responsible for the setup of virtual channels and broadcast and multicast—especially Distant Vector Multicast Routing Protocol version 3 (DVMRP v3)—setup. In order to do so, they must parse the multicast setup messages that determine which ports and IP addresses will be the target for the multicast. The line cards will set up the multicast such that the replication occurs as far as possible to the egress side. So that they do not overload the infrastructure or the node itself, they should be aware of the remaining available bandwidth on any resource they utilize for the multicast or broadcast. The line cards will have to determine both the remaining bandwidth and the stage at which replication will have to occur, and they will have to communicate with external and internal resources to achieve this.

INTERIOR AND EXTERIOR BORDER GATEWAY PROTOCOLS

The line cards will have to be able to communicate with other line cards to exchange routing table information and other information pertaining to forwarding datagrams. These protocols run on top of IPv4 and IPv6, and they require that the line card can parse datagrams for these messages, commands, and status data. The line card must be able to determine when to terminate datagrams and when to forward them. A special case exists when the router, as a single managed device, is the destination of the messages or commands, but each line card individually is the recipient of these messages or commands. The issue is that instead of a single managed entity, the router consists of multiple independent entities, and so not all signaling messages are going to one centralized managed entity, but to multiples of them. In that case, each line card must parse the datagrams, interpret them, extract the vital information pertaining to EGP, BGP, CIDR, or RIP, and then forward the replicated messages and commands through the switch fabric to the other line cards. In any case, the line cards must be able recognize these messages, filter them out, replicate them, and then send them to the other recipients. This can be done through multicast or subaddressing.

In some cases, frame information is required for further processing, and so the framer must not remove all information, but instead retain some of the encapsulating information and embed it in the message. The frame information may be required for any of the forwarding protocols, and that information may include, but is not limited to, MAC addresses and IP addresses of the source and the destination side. In that case, the network processor on the line card must inform the port card's MAC and framer how to proceed with the framing information and address information.

Also, it is important to remember that the datagram must be assembled for the entire message to be available. Trying to extract the message information from the fragments is not possible after segmentation has occurred.

LINE CARD FUNCTIONS FOR PSTN INTERNETWORKING

For the PSTN Internetworking, the line card must be able to not only physically connect to the PSTN through a port card, but it more importantly must be able to understand CCS#7, MFCR2, and all other signaling related to the PSTN and its signaling methods. Communication with the Class 4 or 5 Central Office switches of the PSTN requires that the line card receives, terminates, sends, and forwards messages, commands, and the status data of CCS#7. Terminating messages, commands, and the status data of CCS#7 means that the line card filters these out of the data stream, parses them, and processes them. In some cases, the line card can take action locally after the message has been received and processed; in some other cases it means that the line must generate a message to another CCS#7 entity to have this entity perform a lookup, and then wait for the result of the lookup to take action upon the physical or virtual connection, a datagram, or a set of datagrams. To do this, the line card requires a processor that can very reliably process these messages, commands, and the status data in near-real-time.

PORT CARD FUNCTIONS

The port or interface card connects the external line physically and logically to the processor card. It contains the line specific logic (LSL) and deals with the LSL protocols. Mostly, the port cards handle links that require some low-level processing of the net payload datagrams in order to be transmitted. They may use coding and associate one set of symbols to another, often longer set with redundancies, and may provide a certain symbol rate based on the channel capacity. Since this often gets confused, here is a very short explanation of the terms.

DEFINITIONS

The **Channel Capacity** is defined as the maximum rate at which a physical medium can transport symbols from the sender to the receiver. By definition, the channel capacity depends on the signaling method and the coding method. The channel capacity is measured in baud and expresses the number of symbols (see Symbol) per second that can be transported. The channel capacity is depending on physical parameters of the line.

A **Symbol** is the basic entity of information that is transmitted. In a large number of systems, a symbol is a bit (binary digit) or a byte. A symbol can be, but does not have to be, digital. It can be analog, or a discrete value other than 2. It can even be a vector instead of a scalar. Multi Level Signaling (MLS) uses a number of values other than two per symbol, and in Quadrature Amplitude Modulation (QAM), a symbol can be seen as a vector.

The **Symbol Rate** is the channel capacity of a link or a transport medium. If a symbol can be expanded into more than one bit, the bitrate over a given channel is higher than the symbol rate. This is typically true for MLS. If a system with MLS uses four different levels, and all statuses and status transitions are allowed, each symbol represents two bits. Therefore, the bitrate can be twice as high as the symbol rate. Therefore, a High Speed Serial Link (HSSL) with MLS can have a symbol rate of 2 Gsymbols/s, and provide a bitrate of 4 Gbit/s.

Coding is the process of associating a set of symbols to another set of symbols. As a result, one set of symbols can be replaced by another set of symbols, individually or in sets. Additionally, coding allows for adding redundancies for error detection and correction. Coding allows for bits to be grouped together into larger engines and supports error detection mechanisms through parity generation, checking, and DC balancing in HSSLs. Typical coding schemes are 8B/10B and 64b/66b.

The net data transfer rate results from the symbol rate multiplied by the number of bits per symbol, divided by the code efficiency.

While coding is most often used on external links, newer routers that deploy HSSLs within the device will use 8B/10B on internal interconnects—such as XAUI and in HSBI-compliant links—to connect subsystems with each other.

NETWORK PROCESSOR AND TRAFFIC MANAGER SOFTWARE IMPLICATIONS

We have discussed the implications of the requirements of the future on the hardware of routers—especially the line cards—in detail. The software implications of IPv6 or any other protocol that implies Service Level Agreements with QoS parameters are equally important. First of all, the software must support the hardware in its functions. More importantly, it must do so without sacrificing throughput, delay variation, latency, and robustness. It is of no use if the hardware by itself is crash-proof, but the software running on it frequently crashes or must reboot due to software faults. This means that the software must be capable of surviving some hardware errors, but more importantly, it must have a flawless stack and heap administration. Garbage collection must be flawless. In case something goes wrong anyway—like spurious hardware faults or nested ISRs that terminate because of run-time issues— a planned shutdown and restart (recovery) must be implemented. However, rolling recoveries must be avoided. All levels of the software for all programmable components on the line cards—network processors, traffic managers, and local control processors—are affected by the requirements for the SLAs, QoS, and availability. This is equally the case for data path software and for control path software.

LOCAL CONTROL PROCESSORS FOR LINE OR PROCESSOR CARDS

One could—and should—argue that on a line card, port card, or processor card, an additional local control processor that is not in the data path does not have a significant cost impact; it provides additional capabilities for local control and bootstrapping, control interface hardware to the OAM&P card, and means for statistic traffic monitoring. It might also contain the watchdog logic for resetting the network processor in case the network processor, the traffic manager, or their software malfunctions and fails to return to normal operation.

A local control processor therefore adds significant benefits to the module, especially increasing the reliability of the module by offloading software from a processor that is in the data path. It not only frees resources on the processor—typically the network processor and the traffic manager—it also makes the function split clearer and therefore supports an increase in reliability. While the cost and PCB area are immediately clear, the software effort is not. Typically, the software development effort for these local control processors is in the order of the hardware development cost of the digital portion of the board—but this is not the whole story. If there is some local control on the line or processor card, then the centralized OAM&P card can be divested of a few functions that do not need to be processed device-globally. As a result, the OAM&P card's software can be reduced in complexity and improved in availability and robustness—without impact on functionality.

Functions that are typically executed on a local control processor on the line card, port card, or processor card are bootstrapping, setup and configuration of the network processor, traffic manager and other support circuitry, statistical traffic monitoring, collection of error statistics and heuristic, and sometimes even more complex traffic analysis of data provided by the PHY, MAC, framer, network processor or the traffic manager, such as: number of routed packets or cells over a preset period of time per priority class; dropped cells within that period of time per priority class; the minimum, maximum or average number of queued cells or packets in the switch fabric per priority class; the minimum, maximum and average duration of datagram traversal through the line card, and datagram traversal times through the network processor or traffic manager per priority class. In more elaborate scenarios, this traffic collection scheme is extended to per-IP address monitoring or any other observable part of the datagram that the network processor can detect. The PHY, MAC, framer, network processor, or the traffic manager can also provide bit error rates and cell or packet error rates. The MAC can report on access violations and errors on the MAC layer, whereas the framer will report frame errors or any other anomalies it finds in frames.

The local control processor for supporting in the collection of the statistic traffic data will retrieve all this traffic data from the individual components in the data path, such as transceivers, PHYs, framers, MACs, SAR engines, packetizers, network processors, and traffic managers, and then forward it to the centralized OAM&P entity. It will then summarize it and provide this data in the form of database entries to the OAM&P entity. With information like this from all systems in the data path of any given datagram, it is fairly simple to prove QoS fulfillment (or non-fulfillment) for true IPv6 compliance.

One other advantage of a local control processor is that it can take over communication with the OAM&P card. This communication typically will run over a higher layer software stack on top of Fast Ethernet. As a result, the network processor and the traffic manager on the line card or the processor card do not have to perform these tasks at all, and the software does not have to contain any components for that particular function, including a TCP/IP stack. This shifts the effort to the local control processor and away from the network processor and traffic manager. Instead, the latter two can communicate over a very simple and fast message-passing interface with the local control processor. This in turn completely separates local control messages from data path functions, and as a result makes intrusion into the router much more unlikely. The network processor does not need an IP address, and cannot be a direct target of an intruder. Its policies are not directly accessible, and drop policies as well as firewall functions cannot easily be perpetrated.

COMPUTE EFFICIENCY

Every transistor's power consumption P is given by the equation

$$P = U * I_{Leakage} + \alpha * C * U_{dd}^2 * f$$

This is true up to a threshold value of the clock frequency at which the transistor stops functioning correctly. It is impossible to scale the clock frequency to infinity and determine the limes $f \to \infty$, so that the constant (clock-independent but process-dependent) part of the equation $U * I_{Leakage}$ proportionally converges to zero. U_{dd} is the logic swing of the signal, and α is the "switching factor" or the proportion of the rise and fall time of the signal versus the total cycle time.

Consequently, the power consumption of an integrated circuit is described by the number of transistors it deploys and the clock frequencies they operate at (see Figure 8.24)—and that is even true if there are multiple clock domains.

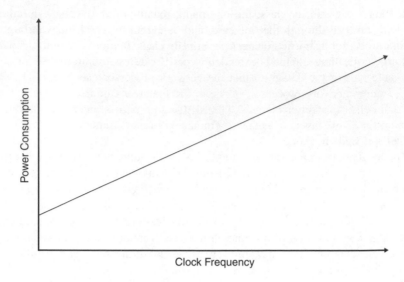

FIGURE 8.24 The Power consumption of an integrated circuit is directly related to the clock frequencies at which they operate.

Gated clocks for units that are only temporarily used or can be shut down when there is no unit activity reduce power consumption, but also reduce the available computational power.

It then becomes crucially important to solve a computational problem in the most efficient way. The availability of transisitors on a die is not the gating element anymore; it is the ability to feed in enough power and to remove the heat effectively.

As a result, it is mandatory to solve a computational problem with the smallest amount of transistors, clocked at a frequency that gives the most efficient tradeoff in terms of leakage current and computational performance.

Moore's Law will still allow for placing more transistors on die. However, it is to benefit from this since it becomes impossible to remove the heat from the die. That is true for multi-core processors as well as single-core processors. The above equation is independent of the number of cores or processors on a die—it only counts transistors. Consequently, multi-core processors are not the solution to the problem of heat removal. They do not compensate for inadequate architectures.

CONCLUSION

Line cards are one of the essential building blocks of a router. They and the switch fabric are the only active components in the data path of the datagrams. The only other component in the data path of all datagrams is the passive backplane or midplane. All other components of a router are either passive, or they are not in the data path. They might not even be in the control path either. Some components are mechanical and do not have any influence over the data path or control path.

As a result, the line cards directly influence the ability to losslessly switch and route datagrams. They parse and process the datagrams, perform lookups, apply policies to datagrams, and perform Segmentation and Reassembly and policing at line speed for thousands of virtual connections simultaneously, thereby enforcing and supporting SLAs, dropping excess traffic, and performing Traffic Management by queuing and buffering traffic according to SLAs. They support Traffic Engineering by means of collecting Statistic Traffic Data, support a mixture of hierarchical and mesh interconnect infrastructure in terms of data traffic, support an overlay network for metadata and signaling data, enable traffic rerouting at the edge, provide redundant fail-safe systems towards the core, and communicate with the PSTN infrastructure. Together with the switch fabric card, they also directly influence many factors: the minimum, maximum, and average Net bit rate; the minimum System availability; the minimum System uptime; the minimum, maximum, and average CDV for ATM networks only; the minimum, maximum and average Logical Connection setup time; the minimum, maximum and average Delay and Latency; and the minimum, maximum and average Round Trip Delay. In other words, it directly impacts the parameters that are measured by SLAs.

The line cards of the new generation of advanced routers also maintain the scalability of the Internet, and therefore truly support the unification of networks.

9 Switch Fabric Cards

OVERVIEW

The purpose of a router is to forward datagrams from an ingress port to an egress port, and to make decisions regarding if and when to discard a datagram. While the line cards (or processor cards in a midplane design) perform the forwarding decision or the drop decision, some entity must connect the line cards to each other. Every router therefore must have an entity that can connect the line cards or port and processor cards with each other through internal interconnects. These interconnects can be buses, meshes, shared memory switches, crosspoint or crossbar switches, or switch fabrics. All of these interconnects differ in their cost structure, throughput, latency, connection setup time, flexibility, resource utilization, and reliability. As a result, they are used in different applications. The requirements for a switch fabric are determined by the requirements for an advanced router.

FUNCTIONAL REQUIREMENTS FOR AN ADVANCED ROUTER'S CORE

Current and future router and switch architectures must cope with a variety of different requirements. One is that traffic continues to grow at a rate of 40–80% annually in the core of the network. Consequently, the switch fabric capacity must be able to grow by at least that factor. Another requirement is that these routers and switches will have to be able to handle IPv6, PSTN TDM, TDM-like traffic, and MPLS to service traffic with a variety of Quality of Service (QoS) parameters while maintaining all Service Level Agreement (SLA) parameters.

As a result, routers and switches designed today, even more than in the past, must be able to support multiple levels of QoS and be field upgradeable. QoS support is required in hardware and software on the line cards, but even more within the switch fabric as the most crucial and central element of a router or switch. Additionally, the switch fabric must be upgradeable in the field without impacting the other components of the router. Furthermore, it must be possible that obsolete line cards can be swapped out without any detrimental effect on the rest of the system.

This requires a multiplicity of features in the core element of a router or a switch: scalability, a clear and clean function split, physical and logical partitioning of the switch fabric components, and full QoS support by means of awareness of priority levels and of median and average delay boundaries.

As we have determined before, the advanced router must be able to perform the following functions and possess a variety of features:

- Losslessly switch and route datagrams
- Be able to perform Segmentation And Reassembly (SAR)
- Perform policing at line speed for thousands of virtual connections simultaneously and thereby enforce and support SLAs
- Perform Traffic Management by queuing and buffering traffic according to SLAs, and drop excess traffic
- Support Traffic Engineering by means of collecting Statistic Traffic Data
- Support a mixture of hierarchical and mesh interconnect infrastructure in terms of data traffic
- Support an overlay network for metadata and signaling data
- Enable traffic rerouting at the edge and provide redundant fail-safe systems towards the core
- Communicate securely within the components of the router
- Communicate securely between the OAM&P card and a Billing Center
- Communicate securely between the OAM&P card and a PKI Center for authentication
- Communicate securely between the OAM&P card and a Network Management Center (NMC)
- Communicate with the PSTN infrastructure

Some of these have a direct impact on the switch core of the router:

- Losslessly switch and route datagrams
- Enforce and support SLAs
- Support Traffic Engineering by means of collecting Statistic Traffic Data
- Enable traffic rerouting at the edge and provide redundant fail-safe systems towards the core
- Communicate securely within the components of the router

It must do all these and fulfill them under the additional challenges of power distribution, space and volume requirements, and the ability to dissipate heat.

HISTORY OF ROUTER-INTERNAL INTERCONNECTS

While in the beginning, shared buses were fast enough to support the external line rates on a multitude of line cards, this was soon not enough anymore. In the early days of router design, a passive backplane with a bus configuration for the line cards was a sufficient solution—the capacity of the internal bus by far overwhelmed the aggregated line card I/O performance. Therefore, no congestion or contention occurred, and performance degradation was not an issue. This has changed dramatically. First of all, the line rates increased dramatically. Second, QoS is now a hard requirement since many carriers have SLAs with their customers, and these include a certain throughput at certain maximum levels of latency for different traffic types at a given service availability. That in turn demands a blocking-free switch entity to handle temporary contention. Shared buses suffer from blocking, congestion, and contention. As a result, a new paradigm was introduced. In this new paradigm, the line cards or port and processor cards were connected to the switch card—and ultimately to each other—via a passive backplane or midplane. In essence, the point-to-multipoint connectivity was displaced by a multiplicity of simultaneous point-to-point connections of all line cards or processor cards to a switch entity on a switch card. The switch card first contained shared memory switches. Shared memory switches quickly ran out of steam; their throughput was limited. Crossbar switches were introduced to replace the shared memory switches, and they provided a truly blocking-free behavior. However, scheduling a large crossbar switch turned out to be quite compute-intensive. Consequently, designers looked for solutions to cope with the scheduling problem while maintaining the crossbar switch as the switching element.

As a consequence, self-routing switches were introduced, based on crosspoint or crossbar switch elements. They were called switch fabrics, and can be described as a crossbar switch with a built-in scheduler for the setup and teardown of connections, and as a result are self-routing. For switch fabrics, the control path is the data path, and therefore a switch fabric must have a parser that uses the control path information to set up a connection through the switch fabric based on this data. These switch fabrics do not require external schedulers or the involvement of the network processor to forward a datagram from ingress to egress. While they were naturally blocking-free, they were not free of contention or Head-of-Line blocking. Consequently, Input-Buffered Output Queued Switch Fabrics were invented. These buffer traffic on their ingress ports and queue it on the egress ports. In addition to the crosspoint of crossbar switch, they provide input buffers and prioritized output queues with built-in schedulers. This made them blocking-free, and together with an internal speedup (higher bandwidth provided internally versus externally) prevented practically all Head-of-Line blocking and eliminated most of the potential internal contention.

As a result, they fit the requirements of router designers, but came at a price. Their memory efficiency is not optimal, and therefore they were an expensive proposition. Very fast on-chip memory is expensive and therefore should be used with the highest possible efficiency. Input-Buffered Output Queued Switch Fabrics did not do that, and as a result, new architectures had to be invented that made better use of the memory. The same advantages of the Input-Buffered Output Queued Switch Fabrics at a much lower required level of on-chip memory came with Virtual Output Queued switch fabrics. Abrizio's Tiny Tera—resulting from a Stanford University project by Nick McKeown—was the first to implement this idea. However, Tiny Tera had a significant drawback. Its queue managers were supposed to be located on the line cards. First of all, most every router designer mistook them as traffic managers and consequently considered them too small to queue and shape traffic. Second, the number of links required to connect all components together was too large. Since the queue managers resided on the line cards, and the crossbar switches were on the switch card, the interconnecting traces had to be routed through the backplane, and more importantly, into the switch fabric card through a connector. There simply were not any connectors in the market that provided enough connections for the implementation of Tiny Tera in a large router. As a result, Tiny Tera was not a huge success. Mindspeed's iScale solved this problem. In iScale, the queue managers were located together with the crossbar switches on the switch fabric card. This ensured that all interconnects between the switch fabric components were routed on a single board and did not have to traverse connectors. Consequently, even a massive internal speedup of two with the consequential increase in the number of interconnecting traces between the queue managers and the crossbar switches did not result in exhaustion of pins in high-density connectors. Additionally, there was no mistaking the queue managers for traffic managers.

This new generation of Virtual Output Queued Switch Fabrics allowed for a new generation of routers that was able to dynamically route and forward packets from multiple line cards each at OC-48 or STM-16 wirespeed or higher with minimal latency and with a high degree of contention resolution and prevention.

Many observers see optical interconnects and full meshes as the possible next steps in router cores. While optical interconnects offer a very high bandwidth per channel, it has not been proven yet that self-routing optical interconnects are feasible. As a result, optical interconnects require external schedulers. Full meshes in the backplane cannot replace current switch cards either, since they shift the effort of switching, routing, and contention resolution from the switch core to the line card. While a full mesh makes the backplane theoretically trivial—but not in reality, since many more traces are required—the line cards under those circumstances have the task of contention resolution.

BASICS

Switch cores come in a wide variety of designs. The easiest and simplest "switch" theoretically is an interconnecting full mesh (see Figure 9.1). It does not require any logic in the "switch" core, and as a result, it does not add complexity or power consumption to the "switch" core. It does, however, add to the complexity of the line card, and it requires an extraordinarily high number of interconnecting traces. In a router with n line cards, each line card must provide $(n-1)$ internal egress and $(n-1)$ internal ingress ports to support the full mesh. This leads to a very large number of interconnecting traces.

The next simplest interconnect is a bus. While all line cards connect to the bus, only one source/destination pair can communicate at any single point in time. As a result, it is blocking. To regulate access, a centralized scheduler or arbiter grants access to the shared resource. However, it uses a fairly small amount of traces, and the scheduler is fairly simple, too. If time division multiplex (TDM) allows for multiple bus transactions per time slot of the external datagram rate, then the effect

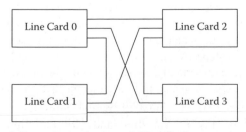

FIGURE 9.1 Full mesh Interconnect between 4 Line Cards.

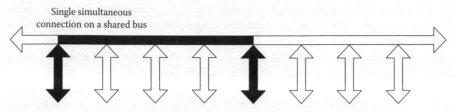

FIGURE 9.2 Shared bus.

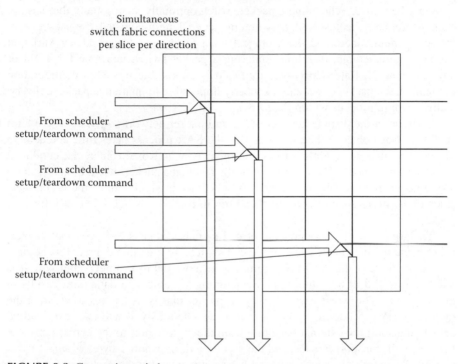

FIGURE 9.3 Crosspoint switch.

of blocking can be minimized and acceptable rates of throughput can be achieved (see Figure 9.2). In case the bus and the scheduler support multicast, one source can transfer datagrams to a multiplicity of destinations.

True switches can be based on a shared memory array, crosspoint switches (see Figure 9.3) or crossbar switches, optical interconnects, and switch fabrics in any buffering and queuing configuration. A switch can be packet-based or cell-based, and it can be externally controlled or internally controlled, and therefore is considered self-routing. Switches can be analog or digital. As a result, they forward a continuum of electrical signals (analog), or discrete—mostly binary— values of voltage. Optical interconnects are the equivalent of analog switches. A crosspoint switch is truly blocking-free, and allows multiple connections at the same time.

Switches can be implemented as packet switches or as cell switches. A packet is a datagram of variable length with the length information either implicitly or explicitly provided. A cell, on the other hand, is a datagram with a fixed length. Since cells have a fixed size, it is fairly simple to switch them without generating any mutual interference within a crossbar switch. For the most part, a cell switch can be operated on fixed time intervals, very much like TDM. In a packet switch, none of this is true. The crossbar switch will either have to accommodate long packets in a TDM fashion, and therefore waste effective bandwidth on shorter packets, or it will block packets that must traverse the same crosspoints that might currently be in use. Scheduling a packet switch optimally is a problem that has not even theoretically been solved. If switching occurs length-adjusted, the packet order cannot be maintained, and the average Packet Delay and Packet Delay Variation become unpredictable, therefore rendering a packet switch unusable for TDM or any Constant Bit Rate (CBR)-type traffic. It is impossible in such a configuration to maintain the packet order, and it is nearly impossible to guarantee packet delivery. Collisions might occur within the switch.

Early on in the dispute between ATM and packet networks such as IP, the term "cell tax" was coined. While it is true that the SAR process takes time and adds to the latency of a datagram's traversal through a switch or a router, the combined effect has little impact. More importantly, the fact that a packet switch is significantly less efficient than a cell switch, and its scheduler is significantly more complex, leads to the preference of cell switches even in routers—which inherently forward packets.

The argument of the opponents of cell switches always has been that packets never come in convenient sizes for the cell switch. To a certain degree, this is true.

Let us look at the "cell tax" argument in more detail to understand why it is valid and invalid at the same time. Segmenting variable-size datagrams results in inefficiencies. The worst case is always a packet that is $n+1$ bytes long when the internal payload of the internal cells is n bytes. Obviously, it will take two internal cells to transport the external datagram from the ingress port to the egress port. The first cell will transport n bytes, and the second will transport 1 byte, with $n-1$ bytes unused. There will always be waste of bandwidth provided the packet size is not an integer multiple of the internal datagram payload size.

The number of cell-sized fragments N_C, and the number of fragments N_F in size between 1 and the internal cell size minus 1, both are a function of the internal cell size L_C and the largest allowable packet size L_P.

From this, we can derive that the internal cell size significantly impacts the efficiency of the line and the switch fabric. The distribution of the internal size of the cells to be forwarded through the switch fabric depends on the fragment length L_F, ranging from 1 to the internal cell size minus 1, and on the number of the fragments having the internal cell size.

If $L_P = L_F + n * L_C$, with $L_F = 0$ or $L_P \bmod L_C = 0$, then the number of fragments N_F, sized from 0 to the internal cell size minus 1, is the modulo of the largest allowable packet size L_P and the internal cell size L_C each, or $N_F = L_P \operatorname{div} L_C$, and the number of fragments having the cell size $N_{F,C}$ is given by $N_{F,C} = \frac{1}{2} * L_P \operatorname{div} L_C * (L_P \operatorname{div} L_C + 1)$. If $L_P = L_F + n * L_C$, with $L_F \neq 0$ or $L_P \bmod L_C \neq 0$, then the number of fragments N_F sized from 0 to L_P mod L_C is the modulo of the largest allowable packet size L_P and the internal cell size L_C each, $N_F = L_P$ div $L_C + 1$, and the number of fragments N_F sized from L_P mod $L_C + 1$ to the internal cell size minus 1 is the modulo of the largest allowable packet size L_P and the internal cell size L_C each, or $N_F = L_P$ div L_C, and the number of fragments having the cell size $N_{F,C}$ is given by $N_{F,C} = ½ * L_P$ div $L_C * (L_P$ div $L_C + 1)$.

In any case, for all packet lengths greater than twice the cell size, the number of cell-sized fragments is larger than the number of fragments for any other size. As a result, in that case the efficiency of a cell switch is greater than 0.5 under all conditions. The larger the packet size, the closer the efficiency is to 1.

The realistic case is that the two maxima are at 64 bytes for all the HTTP put commands and 512 bytes packet length for any file transfers, and they taper off very quickly after that. As a result, the efficiency is skewed even more towards 1 because of a further predominance of the 64-byte fragments and cells. We can therefore assume that the efficiency of a cell-based line card and switch fabric is greater than 0.5 and less than 1.0—but it is closer to 1.0 than to 0.5.

This is a huge improvement over packet-based line cards and switch fabrics, where the efficiency becomes smaller when the distribution range of the packet fragment sizes gets larger. The efficiency of these packet-based line cards and crossbar switches is significantly below 0.5, if not below 0.10.

The effect of lost time slots because of switching packets with variable lengths is much more pronounced. Not only is the scheduler significantly more complex, the crossbar cannot switch on fixed-length intervals. It must switch-based on the maximum length of a packet in transit through the crossbar, and renders all other timeslots useless, even if all other packets were small and have finished traversing the crossbar. As a result the scheduler will have to track the traversal times of packets through the crossbar, and schedule connections based on that.

As we can see, the crosspoint or crossbar switch is very straightforward and easy to handle, especially if it is a cell switch. The only remaining challenge is contention on egress and Head-of-Line blocking, and the process of scheduling itself.

Head-of-Line blocking can to a large degree be prevented with a combination of Virtual Output Queus—which are on the ingress side—and internal bandwidth overprovisioning between the queue managers and the crossbar switch. Output Queuing can reduce contention on the egress link of the switch fabric into the egress-side traffic manager or network processor. The process of scheduling the switch fabric and the queue managers is offloaded from the network processor and therefore does not require any resources there.

In any case, the backplane or midplane is passive. There must not be any active components on the backplane—not even electrolytic capacitors. Ceramic capacitors and resistors for termination of lines are the only allowed passive components on the backplane or midplane. All active components must be on the cards that plug into the backplane or midplane or other removable subsystems of the router. The switch core and any line cards plug into the backplane or midplane and connect the line cards to the switch core.

EXTERNALLY CONTROLLED SWITCHES

Externally controlled switches are switches such as crossbar switches or crosspoint switches that require an external entity to set up a connection and tear the connection down. Optical crossconnects are other examples of externally controlled switches. The connection setup and teardown is initiated by an entity external to the switch.

Externally controlled switches are based on unbuffered and unqueued crosspoint or crossbar switches. In externally controlled switches, the control path is separate from the data path. A connection setup or teardown command is first sent to the scheduler through the control path, and the connection is set up. A setup completion message is sent back through the control path, and the network processor or traffic manager can then send the net user data through the data path to the switch. As a result, the network processor or traffic manager must buffer or queue the datagrams destined for the unbuffered switch as long as the switch cannot accept them. Teardown can either be automatic for cell-based switches or initiated through commands for packet-based switches.

As a result, externally controlled switches are not too well-suited to dynamic switching. They do quite well in environments with quasi-static connections. Whenever dynamic switching is a necessity—which is the case in most routers—then a simple crosspoint switch with an external scheduler (see Figure 9.4) or an arbiter is not sufficient. For a variety of reasons the scheduling of packets cannot be accomplished in the available time frame. Variable packet lengths prohibit blocking-free operation, and so the packets must be segmented into cells of fixed size such that the switch can essentially switch on a TDM basis and switch cells on a fixed timing interval. To achieve this, the line cards must segment packets into cells and add header information such that the priority, the destination port, and the cell sequence number are included in the cell header information. This is what the ingress-side line cards will have to perform to ensure the switch is blocking-free. On egress, the line cards will have to reassemble the cells into packets. Consequently, the reduction of complexity in the externally controlled switch is traded, versus an increase of complexity in all line cards. The process that is involved with this is described in more detail in the Line Card chapter of this book. Switching packets instead of cells requires a very high scheduling performance.

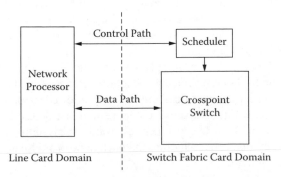

FIGURE 9.4 Crosspoint switch with external scheduler.

THE SCHEDULING CHALLENGE

Currently, most routers deploy buses or unbuffered and non-queued switches to interconnect line cards. These are not self-routing; they need sophisticated arbitration units or schedulers to ensure that there are no conflicting data transfers over the shared bus. These schedulers and arbitration units are part of the line card, its network processor, or its software. Newer routers deploy self-routing switch fabrics; some of them use Virtual Output Queued switch fabrics. For self-routing switch fabrics, the process of scheduling the switch fabric and the queue managers is offloaded from the network processor and therefore does not require any resources there.

A switch with n input ports and n output ports has n^2 interconnection points. Any n ports can be interconnected at the same time provided that there is no conflict on ingress or egress (contention). The maximum number of connections that can be set up per direction is n. In both directions combined, that number is $2n$. That would assume that all connections could be made without collisions. Mostly, it is to be assumed that only $n/2$ connection setup requests can be issued at once. Therefore an internal speedup of a factor of two would allow all possible connections to be made within one cell cycle. The number of connections set up and torn down per cell or packet time nevertheless is quite high, and so the appropriate protocol becomes an important issue. This means that in order to not lose any time slots available for being switched, the communication between the switch core and the queue manager must be fast. While this is a large amount of bandwidth required just for scheduling the connection setup and teardown, it is manageable in a single-stage switch fabric. In a multi-stage switch fabric, the number of interconnections increases quadratically. This is surely not possible with external schedulers. While it might be possible to increase the combined bandwidth through multiple stages of switch fabrics or crosspoint switches, the scheduling performance does not keep up, and it becomes the bottleneck.

CROSSPOINT SWITCHES AND CROSSBAR SWITCHES

Crosspoint or crossbar switches are switches that are based on (mostly) analog switches in a crossbar pattern that can connect any one input to any one output at a time. Since by definition they are non-blocking, any one or more of multiple inputs can be connected to any one or more of multiple outputs, for as long as there are no inputs or outputs that are sources or targets of multiple connections at the same time.

SELF-ROUTING SWITCHES

Self-routing switches have a built-in scheduler that allows for autonomous connection setup and teardown. Shared Memory Switches and Input-Buffered Output Queued Switches as well as (combined) Virtual Output Queued Switches fall in this category. Self-routing switches are characterized by a scheduler used to set up connections and tear down connections based on the header information of a datagram. As a result, the network processor does not have to be involved in the connection setup or teardown. While this may not appear to be a huge challenge, clearly distributed processing and scheduling is necessary to fulfill the requirements for line rate scheduling performance. Distributed scheduling allows linear growth of a switch fabric. A centralized scheduler in switch fabric architecture basically prevents any linear growth of throughput. To prove this, let us perform a few basic calculations for a hypothetical switch fabric with a centralized scheduler.

10 Gbit/s per port translates to 31.25 million packets per second (MPPS) in worst-case scenarios at 40 bytes per packet, even if the switch fabric is cell-based and uses a 64-byte internal net payload datagram size. With port overspeed in applications with a traffic manager, it can easily add 50% to the arrival rate of datagrams, which in turn translates into 46.875 MPPS per 10 Gbit/s port. For OC-48 systems, these numbers are 7.813 MPPS and 11.719 MPPS per port, respectively. A typical 16-port system therefore would have to handle 16 multiplied by 11.719 MPPS for OC-48 line speeds or 16 multiplied by 46.875 MPPS for OC-192 or 10GbE systems. This is 187.5 MPPS in total for OC-48 systems, and 750 MPPS for OC-192 and 10GbE systems.

Very obviously, these numbers are huge. Even if these enormous amounts of scheduling decisions can be made, it is impossible to transfer these setup and teardown operations from a centralized scheduler into a switch fabric. Therefore, the scheduling decisions need to be made where the queuing occurs and where data enters the switch fabric so that the parsing engine can determine egress port number and priority on a per-port basis.

For self-routing switches, the data path physically is the control path. In other words, the datagrams sent by the network processor to the switch fabric contain a local header—the LCI—and the payload data. Typically, a header is prepended to the net user data payload. In most cases, self-routing switches switch fixed-length datagrams, with fixed-length payload data and fixed-length headers. They perform cell-based switching. All self-routing switches require a local tag in order to switch the datagram. This tag is most often referred to as LCI, or Local Connection Identifier.

LCI

The Local Connection Identifier (LCI) must contain at least the destination port address. In practically all cases in a modern router and therefore in a switch fabric, it will have to contain the priority of the datagram, or any other descriptor for its requirements regarding real-time forwarding. Aging descriptors are required for the traffic manager. If Reassembly on egress is required and no other means of identifying the ingress port are provided, then the LCI will have to contain the ingress port number. If there is any possibility of internal datagrams arriving out of sequence, then the LCI will also have to contain a sequence number that runs at least deep enough to identify internal datagrams as out of sequence and possibly even to repair out-of-sequence arrivals of internal datagrams. The switch fabric only needs the destination port address and the priority of the datagram. However, all these descriptors are typically contained in the LCI; therefore, they increase the local overhead and decrease the net payload efficiency.

SHARED MEMORY SWITCHES

Shared Memory Switches are based on arrays of dual-ported memory cells that are accessible through a centralized scheduler. The individual memory cells allow quasi-simultaneous write and read accesses by subdividing an external cycle into a write cycle to memory, a command cycle, a status cycle, and a read cycle from memory. The row and column addresses are used for accessing the cells, with the row and column address being associated with the ingress port/egress port tuple. The scheduler typically allows for multicast and broadcast transmittals, and must resolve contention.

Some proponents of shared memory switches claim that they are not adversely affected by multicast traffic. This is not true. Multicast traffic is characterized by the fact that it inherently has a 1 to n traffic fanout. In effect, it overloads egress ports because it can and will interfere with other multicast or unicast traffic targeting the same egress port. It shares this characteristic feature with all switches. Although it is true that shared memory switches write the datagram only once into memory and read it out multiple times, this is not an advantage over any other switch that is capable of replicating the datagrams that are to be multicast or broadcast. In fact, most switches are capable of replicating the datagrams for multicast instead of sending them multiple times internally. In reality, shared memory switches are currently at their limits because they require multiported memories internally for data consistency running at four times the clock rate of a crosspoint switch.

NON-BUFFERED, NON-QUEUED SWITCH FABRICS

Non-buffered and non-queued switch fabrics are crosspoint switches that have a built-in scheduler. In effect, they are crosspoint switches with schedulers that are either integrated onto the die (and therefore are only colocated, but not logically integrated) or are integrated logically into the crosspoint switch. The latter is much more powerful since the scheduling is part of the connection setup and teardown logic. They are often used as basic building blocks for VOQ or CVOQ switch fabrics.

BUFFERED AND QUEUED SWITCH FABRICS

Buffered and queued switch fabrics come in two distinctly different implementations. The predominant architecture provides buffers or queues on a per-row and per-column basis, on ingress or egress, or on both. The other architecture makes buffers or queues available at each crosspoint between rows and columns. Obviously, one implementation requires $2n$ buffer or queue sets with schedulers, whereas the other requires n^2 buffer or queue sets and schedulers per direction. While it intuitively seems better to use the approach of a buffer or queue set at each intersecting point, the memory requirements make this prohibitively expensive. In a 16-port buffered or queued switch fabric, the first approach will require 32 sets of buffers or queues per direction, while the second one will need 256 sets of buffers or queues per direction. Even if the buffers or queues at the intersections perform virtual output queuing and output queuing, and the other architecture requires an additional set of queues for output queuing, then the ratio still is 64 sets versus 256 sets. As a result, on the same die size, the virtual output queues and the output queues of the architecture with per-column and per-row queuing can have four times the capacity of the architecture with per crossconnection queues. That would typically render a much better Head-of-Line blocking prevention and a much better contention resolution.

COMBINED VIRTUALLY OUTPUT QUEUED (CVOQ) SWITCH FABRICS

Combined Virtually Output Queued (CVOQ) switch fabrics provide virtual output queues on ingress and output queues on egress. Virtual Output Queues (VOQ) are located on ingress, but enqueuing and dequeuing is performed on a per-egress port and per-priority basis. Under all reasonable assumptions for traffic distribution, CVOQ switch fabrics provide the highest throughput at the lowest levels of latency and provide the lowest probability for Head-of-Line blocking and contention. The virtual output queues and the output queues and their schedulers can be integrated with the crosspoint or crossbar switch or they can be located in separate Integrated Circuits. In most cases, the queues and their schedulers are located in a separate IC, and these ICs are typically called queue managers. Those queue managers can be intended to be physically colocated with the crossbar switch, or they can be placed away from the crossbar switch. In the earlier designs, they were located on the line cards. Later architectures placed the queue managers on the switch fabric cards.

CLEANER AND EASIER LOGICAL AND FUNCTIONAL PARTITIONING

First of all, it must be mentioned that with the queue manager on the switch fabric card, the function split and the partitioning of functions in hardware and in software is cleaner than with any other existing or planned architecture. The line card deals with all line and protocol-specific issues, as there are internal and external protocols, handshake procedures, messages and commands; it deals with the specifics of the data, including its burstiness and its distribution over time. The switch fabric card, however, is completely independent of this and therefore is unaware and agnostic to the specifics of the data traffic on the exterior. Its job is to switch and to prevent blocking, Head-Of-Line blocking, and any other contention except congestion management. Congestion management is a line card issue again because congestion occurs on the uplink, on egress. If the queue manager is on the line card, then it becomes undistinguishable for the hardware and the software designer which problems are caused by contention, and which problems are caused by congestion. This then leads to confusion on the implementer's side and therefore will cause problems in operation. If the traffic management occurs on the line card, and contention management occurs in the queue manager on the switch fabric card, this confusion does not even arise. Building the hardware and writing the software becomes easier with a clear function split and results in fewer hardware and software interface issues between the logical building blocks. This in turn will not only speed up the development of the system, it will also help in reducing software errors and containing them in case they occur. In other words, processing and conditioning of the datagram is performed on the line card, and switching is done on the switch card. Ideally, the line card should only be aware of the fact that there are egress line cards, but it should not have to handle the issue of switching datagrams or forwarding them to the egress destination at all.

EASIER ROUTING OF TRACES ON THE BACKPLANE

The placement of the queue manager on the switch fabric card(s) also allows for easier routing of traces between the line cards and the switch fabric card(s) over the backplane, and therefore makes the layout of the backplane significantly easier. With the queue manager on the switch fabric card(s), the layout of the traces on the backplane resembles a star, and therefore is not only easy, it also reduces cross talk between signals and prevents traces from crossing over or intersecting. With the queue manager on the line cards, however, the mesh on the backplane becomes non-routable, and interference and cross talk degrade signal quality. It is not a star any more, and in fact it will become a mesh between stars. This requires more layers on the backplane, or it will enforce signal traces to cross each other or intersect. Both are highly undesirable, especially if High Speed Serial Links (HSSLs) are deployed. The problem becomes even more dramatic if redundant switch planes are desired, because parallelizing the two planes of the switch requires the meshed stars to be connected to both planes. In case the queue manager is on the switch fabric card, the line cards are connected to each plane in a star configuration, and if the switch planes are in the center, the topology is an overlay of two stars. Even this is comparably easy to route and to lay out. In addition to that, the queue manager on the switch fabric card allows for the internal overspeed (bandwidth overprovisioning) between the queue manager and the crosspoint or crossbar switch to fully be utilized and implemented with the appropriate number of traces; because of the shorter interconnects, even HSSLs without encoding but with forwarded clocks can be used. In case the queue manager is placed on the line card, the internal overspeed is either reduced to the port overspeed, or there is a requirement of routing an appropriate number of traces through the backplane. Both of these are highly undesirable.

HIGHER THROUGHPUT OF THE SWITCH FABRIC

Placing the queue manager on the switch fabric card results in lower round-trip delays between the queue manager and the crosspoint or crossbar switch. This is not only due to distance between the chips and the speed at which the signal travels (around $3 * 10^8$ ms^{-1} or 30cm/ns), it also allows for using links between the queue manager and the crosspoint or crossbar switch that have a lower latency than either traditional HSSLs with embedded clocks or other interconnects that can link line card components with switch fabric card components. Every 30cm of distance that can be omitted between the queue manager and the crosspoint or crossbar switch allows for one more nanosecond to be used for scheduling just because of the reduced travel time; in addition to that, shorter distances allow for the use of clock-forwarded links without the need to impose any encoding on the signal. Additionally, serial-to-parallel conversion and vice versa can be reduced to the absolute necessary minimum, again reducing latency. The lower the latency, the shorter the scheduling pipeline for scheduling a connection setup in the crosspoint or crossbar switch, and in turn the shorter the delay between cell arrival and cell departure. Cell Delay and Cell Delay Variation (CDV) both benefit from this. Lower CDV enables the switch fabric to be used to more easily switch synchronous traffic. Switch fabrics with higher CDVs need deeper buffers to realign synchronous traffic and therefore impose an additional delay.

This again is easier to do if the queue manager is located on the switch fabric card and if the traffic management and shaping is performed on the line card.

Less Incremental Cost for Upgrades

There is also a cost issue associated with the placement of the queue manager on the line card. Since there are many more line cards in a system than there are switch fabric cards (typically eight to sixteen line cards for any one or two switch fabric cards), and systems are sold with underpopulated line cards or empty slots for line cards, each and every additional line card will have to hold a queue manager, and therefore is more expensive than a line card that does not have to contain a queue manager. In other words, if the queue managers are located on the line cards, it will cut into the margin of the system manufacturer of the router, the switch, or any other node. Since the average selling price (ASP) of a line card is limited, every additional cost of the line card to the system manufacturer will decrease its margin, and therefore reduce the overall profit. This is not true for the basic system consisting of the chassis, the backplane, the power supplies and the fans, the OAM&P cards and the switch fabric cards. Here, scalability will pay off, and the base price is of less importance. A system that truly can scale up will achieve a higher ASP on the market due to its built-in futureproofness. The buyer knows he will incur smaller incremental cost if he upgrades and extends a scalable system, and therefore puts an emphasis on this. The line cards themselves also benefit from the fact that they are freed of providing space and power to switch fabric components. Since they are densely populated with packet- and frame-processing components anyway, every single component they do not have to incorporate is a benefit for the overall design. In addition, they do not have to carry components that dissipate heat, thereby contributing to thermal problems. One other issue occurs if the port speed of the switch fabric does not match the line speed of the port card. In these cases the data is first routed to the line card with the queue manager on board, and from there it is routed to line cards or port cards that do not incorporate a queue manager—again mandating one more type of line card and contributing to a huge unnecessary variety of line cards.

METRICS OF SWITCH FABRICS

Switch Fabrics can be judged based on a few parameters, aside from the obvious number of ports and the nominal data rate on each of the ports. These parameters are as follows:

- Net Bit Rate or Link Rate Utilization
- Throughput

- System Availability
- System Uptime
- Reliability
- Logical Connection Setup Time
- Logical Connection Teardown Time
- Delay and Latency
- Round-Trip Delay
- Cell Delay Variation
- Scalability
- Field Upgradability
- Resource Utilization on the Network Processor
- Cost Structure
- Feasibility

Once these parameters are known, an informed and educated decision can be made as to which switch fabric is right for the project. In some cases, all of these parameters are required to make a decision, while in other situations a subset of these is sufficient to make a decision. However, it should be stressed explicitly that two parameters are often underappreciated: the Net Bit Rate or Link Rate Utilization, and the Resource Utilization on the Network Processor. Both of these are often ignored until it is too late.

NET BIT RATE OR LINK RATE UTILIZATION (MINIMUM, MAXIMUM, AVERAGE)

Why is it important that the switch fabric can achieve 1 Erlang of traffic on each egress port? While wavelengths are cheap, capacity over these wavelengths is not. That means that the advanced router must be able to load the egress ports to the maximum possible extent. Since it is not in control of the ingress ports, it must try to load the egress ports as much as it can. That typically means that traffic will have to be queued if the combined arrival rate of traffic exceeds the combined departure rate of traffic due to contention on one or more egress ports. While it appears counterintuitive that the combined arrival rate can exceed the combined departure rate of traffic, it is not. Even if each port has the same capacity in ingress and egress directions, the arriving traffic may have a temporarily uneven statistical distribution of its destination port and priority tuple, which might very well mean that there is contention on an egress port on a per-priority basis. While statistically, the traffic is evenly distributed, this is not necessarily true at each and every moment. That means that two or more arriving datagrams target the same egress port, and therefore will have to contend for capacity on that port. This is not avoidable without connection admission control (CAC). As a result, the switch fabric will have to cope with these variations of evenly distributed traffic. The larger the distribution variations of the destination ports, the larger the internal speedup and the queue capacity of the switch fabric. However, it is not really a solution to provide an infinite internal speedup (overspeed) or infinite queue depths. The first will lead to overly wide buses or HSSL link lanes between the queue manager and the crossbar switch itself, and the second will lead to not only huge queue managers and queue memory, but also to a very large latency in travel time of datagrams through the switch fabric. Neither one is acceptable.

Higher performance and throughput of switch fabric will allow for better link saturation due to reduced round-trip delay between the switch fabric components. This is especially important in environments where the uplink is using a WAN link. A leased line with OC-192 speed rating allows a data throughput of 9.952 Gbit/s at a cost of about $300,000 per month. At 50%, line rate utilization therefore provides an effective throughput of 4.976 Gbit/s, enough for 88,857 dial-up customers with 56k modems. However, if the link rate utilization approaches 100%, at the same cost for the provider the same link will carry twice as many customers. Thus, the $300,000 can be evenly distributed over 177,714 users instead of over 88,857. Therefore, the basic cost point per user is only half of the scenario with the 50% saturation without any penalty in quality or availability. Providers using a router that can saturate any given link better than another will have an immediate return on its investment. A router that is more expensive by $300,000, but can load the link to its maximum, pays off within two months. A router can do so if the switch fabric card supports it and if the line card assists in shaping traffic.

THROUGHPUT (TOTAL AND ON A PER-LINK BASIS)

Throughput of a switch fabric is crucial to the performance of a router. As a result, the throughput of a switch fabric, under the given circumstances, determines the throughput of the router. The throughput depends on the ability of the switch fabric to dynamically switch cells, and on the rate at which it can load (saturate) the links. In other words, the total throughput depends on the ability of the switch fabric to load each individual link. This is true especially for the egress links, and in order to achieve the highest possible total throughput, the switch fabric must be able to load each egress link. This can only be achieved if all ingress links are loaded to 1 Erlang and the traffic is distributed in a Gauss- or Poisson distribution with regard to its destination port and priority.

SYSTEM AVAILABILITY (MINIMUM)

A high degree of system availability through redundancy and fail-safe operation can only be achieved by architecture and proper design. An incapable architecture cannot be retrofitted to provide fail-safe operations and redundancy. Therefore, the architecture must be set up such that it supports a fail-safe mode and independent operation of the switch fabric planes. This mandates that there are no single points of failure, but it also mandates that there are scheduling engines on each plane that work independently of each other. It also means that it is not advisable to synchronize the switch fabric planes. Synchronizing them renders both unusable if the synchronization method fails—which typically is the case if and when one of the planes is taken out of service. In other words, a core system architecture must be fail-safe and fault-tolerant. An error that occurs while data is in transition, processed, stored, or queued must not cause an action that impacts the integrity of the system or the data transmission for other datagrams. It must be set up such that active and hot standby units both perform the same task independent of each other. Live data is sent to both and received from both, and the receiving entity selects, based on OAM&P input, which of the forwarded datagrams to process further. In a system that is redundant by these means, hot swap, switchover, in-service upgrades, and other maintenance operations do not disrupt the traffic significantly nor impact system integrity. OAM&P cannot predict HSSL datagram boundaries, and therefore cannot losslessly switch over. At least one datagram will be destroyed. Moreover, the condition for switch over in a non-recoverable fault condition is that non-recoverable errors have occurred. That means that datagrams have already been lost, and that they could not be reconstructed or in any other way repaired. A lossless switchover therefore is a myth; detecting errors implies that errors have occurred and it is impossible to go back in time and fix past erroneous cells. Therefore, having detected errors on one link set—and determined that the other link set has lower bit error rates—mandates a switchover, but it does not have to be lossless since data already has been lost. The only mandate here is that the switchover should occur without introducing a significant number of new erroneous (non-recoverable faulty) cells.

System availability is an important item for carriers and all backbone network operators. While carriers are contemplating using datacom equipment for their needs, especially in the backbone, they refrain from doing so due to fear for lack of system availability. Current router architectures have many challenges. First of all, they must allow the system manufacturer to make use of the economies of scale, especially for line cards. They also have to enable the Internet Service Providers to offer and support SLAs in order to be profitable with their services. Therefore, they must provide High Availability—typically "five nines" or "six nines" (99.999 or 99.9999%) system availability at full support for QoS. High Availability requires redundant fail-safe core components, and the switch fabric is the core component in a router. In a crosspoint or crossbar switch-based router, the core includes the switch and the scheduler. In VOQ or CVOQ switch fabric-based architectures, the queue managers are part of the core of the switch, and as such must be included in the redundancy protection. If the queue managers are placed on the line cards, they are part of the non-redundant non-core part of the router, and therefore they are not replicated. There is no protection switching possible for queue managers placed on the line cards. If, however, the queue managers are placed on the switch fabric card, they are part of a core portion of the router, and therefore they are part of a redundant subsystem. If the subsystem or any part of it fails, there is a redundant hot-standby card, and therefore a switchover, in case of a necessary protection-switching operation, will circumvent the problem in its entirety. This switchover will enable the previously hot-standby switch fabric card and disable the previously active switch fabric card with all its components, including the defective queue manager. The resulting service disruption to the line cards is minimal. The defective switch fabric card can be swapped out (hot swap is possible), and again a spare is available. This is an important issue; because the queue manager is a very complex chip, its failure probability is not low. A queue manager placed on a line card cannot be hot swapped. If it must be swapped, the line card must be exchanged. This means a much longer service interruption because the line card holds all line-specific data, including routing tables. These must be reestablished once the card is plugged back in. But swapping a line card also means detaching all its external connections, and reattaching them to the new card after a replacement card is put in place. Therefore, swapping a line card takes more time than replacing a defective switch fabric card from a redundant pair. In other words, if the queue managers are located on the switch fabric card, then they are included in the redundancy architecture. If they are placed on the line cards, they are not a part of the core and therefore are not included in the redundancy architecture. A failure of the queue manager on the line card may affect not only the line card it is mounted on; it might also affect other line cards. Higher system availability therefore requires that all switch fabric components, as core components of the router, are physically located on the switch fabric card. The queue manager on the switch fabric card provides significantly higher system availability since all switch fabric components are covered by the core component redundancy that way. If the queue manager were on the line card, then a queue manager out of service will not only affect the line card, but will also affect all associated VOQs, and therefore switching data to and from other line cards. If the queue manager is located on the line card, it cannot be protected by the redundancy scheme of the switch fabric card. Since the switch fabric card is a centralized resource, it is covered by the redundancy scheme. Excluding the queue manager from this has an impact on system availability. The switch fabric card can consist of components that are load-sharing and act as distributed processing elements. Consequently, a centralized resource by itself does not exclude distributed processing. The opposite is true: if the components are in close proximity, distributed processing becomes more feasible, since sharing tasks does not incur performance losses from latency imposed by transferring data and context of the task. All systems with queue managers on the switch fabric cards benefit from their internal architecture of distributed scheduling and low latency interconnects of the components. This would not be possible with router architectures that place the queue manager on the line cards.

SYSTEM UPTIME (MINIMUM)

The system uptime is mostly determined by the mean time between failure (MTBF) of any subsystem and the ability of the system to cope with the outage of one subsystem. In a redundant system, uptime is not affected by the outage of one part of a redundant subsystem. The system is only affected if no redundant modules are available anymore to take over for the failed parts. The switchover is mostly determined by the strategy to declare an outage and by the setting of the redundant part as hot-standby or cold-standby. The OAM&P entity is responsible for those, and as a result has a big impact on the system uptime in routers with redundant switch fabric cards. Since the switch fabric is a core subsystem, it is advisable to set up the switch fabric subsystem as a redundant pair of switch fabric cards, with an active card and a hot-standby card. This way, the router is available for switching and routing traffic, provided at least one functioning switch fabric card is available. System uptime often refers to the availability of the system for commands and messages, and does not necessarily mean availability for switching and routing user traffic.

RELIABILITY (ERROR RATE)

The reliability of a switch fabric is determined by its internal architecture and by its I/O links. While the internal crosspoint or crossbar switch should be lossless, scheduling may contribute to a certain internal loss of datagrams. This is not very desirable, but is to some extent unavoidable. The same is true for buffering and queuing. The memory should be lossless, but it cannot be guaranteed that the schedulers are always able to enqueue and dequeue losslessly at the same time. However, those error rates should be significantly better than the combined expected error rates on the I/O links. In most cases, modern switch fabrics deploy High Speed Serial Links (HSSL). These links operate with an expected error rate better than 10^{-13}, sometimes better than 10^{-15}. However, due to the fact that a switch fabric typically supports a large number of ports, and those ports typically have multiple HSSLs per port, the error rate is a compound rate. Additionally, the HSSLs' data rate is in the area of 10^{10} bit/s, and so it only takes 10^5 seconds (around 3 hours) to reach 10^{15} bit transferred. As a result, the HSSL, and therefore the switch fabric, must be able to detect practically all bit errors. This is achieved mostly by intelligent encoding (most commonly used are 8B/10B and 64b/66b) of the bitstream and by securing the bitstream with higher layers of error checking, such as Forward Error Correction (FEC). Undetected bit errors in the payload may be a nuisance, but they corrupt only one cell and therefore only one packet. Undetected bit errors in the header will misdirect a cell and therefore destroy a packet that cannot be reassembled due to one missing cell, and also another unrelated packet where an unexpected cell arrives that does not belong in that particular reassembly bucket.

LOGICAL CONNECTION SETUP TIME (MINIMUM, MAXIMUM, AVERAGE)

The logical connection setup time depends on the quality of the scheduling algorithm in the Input Queue or Virtual Output Queue, the quality of the scheduling algorithm in the crosspoint or crossbar switch, the quality of the scheduling algorithm in the Output Queue, and the datagram type. In a cell-based switch, the connection must be held only as long as the cell is in transit, and any length information can be configurable or hard-coded. In a packet switch, the situation is much more complex, and the connection hold time depends on the packet length. The important parameters for the minimum, the maximum, and the average logical connection setup time determine the quality of the switch fabric.

LOGICAL CONNECTION TEARDOWN TIME (MINIMUM, MAXIMUM, AVERAGE)

The logical connection teardown time depends mostly on the datagram type. In a cell-based switch, the connection must be held only as long as the cell is in transit, and any length information can be configurable or hard-coded. As a result, in a cell switch the teardown can be automatic, and the entire switch can be operated in a TDM fashion. In a packet switch, the situation is significantly more complex, and the connection hold time depends on the packet length. More importantly, the longest packet in transit will block the switch and its schedulers from scheduling new connections, even if bandwidth on other crossconnects may be available. The important parameters for the minimum, the maximum, and the average logical connection teardown time determine the quality of the switch fabric.

DELAY AND LATENCY

The latency t_D of a cell switched through a switch fabric is determined by a variety of parameters. This latency is defined as the period of time it takes for any given cell from the incoming port to be switched through the switch fabric and arrive at the outgoing port.

These contributing delays are the travel time (propagation delay time t_p) through all involved chips, the time required by the scheduling algorithm t_s to schedule the incoming cell to be switched, and the time the switch requires to set up the internal connection and switch the path t_c. If there are serialization and deserialization processes required, then they contribute to the latency, too. Their influence is $t_{S->P}$ and $t_{P->S}$, respectively. It depends on the number of these processes involved, the word length, and the link speed. The delay is applicable to each individual component in the data path within the switch fabric.

The total latency is $t_D = t_p + t_s + t_c + t_{S->P} + t_{P->S}$.

It can be broken down into several parts, and some portions are constant, whereas others are easily influenced by the chip and system designer.

The time required by the scheduling algorithm t_s to schedule the incoming cell depends on a multitude of parameters. It depends on the type and implementation of the algorithm, of course, but also depends on the number of ports n and the load of the links—the link rate utilization R in Erlang. Obviously, the number of ports n linearly influences the latency, even if the switch fabric is blocking-free and Head-of-Line blocking-free. Also, it depends on the internal speedup (bandwidth overprovisioning) of the link between the queue manager and the switch element, on the fill level of the queue, and on the availability of buffer space in the egress-side queue manager. It is dependent on the ability of the scheduler to schedule a path between the ingress-side queue manager and the switch element, and on the switch element and the egress-side queue manager. The datagram traversal time heavily depends upon the latency of the protocol between the individual chips, especially in high load conditions; the datagram must be scheduled for transmission when a link, a time slot, or any other type of capacity for the basic datagram size entity is available.

The design of the switch fabric can influence the implications of some of these parameters. Some are fixed and constant. This is true for the propagation delay time t_p because it is determined by the speed of light or the speed at which electrons move. There is not a lot designers can do, except to make the paths shorter.

However, if the switch fabric is split up into the queue manager and the switch element for ingress and egress buffering and queuing, these two components must be located as close as possible to each other in order to enable the handshake between both components to run without any delay. Any delay in the Request/Grant handshake protocol will delay the decision of how and when to set up a path from the ingress-side queue manager through the switch element into the egress-side queue manager. As a consequence, and if that cannot be finished within one cycle, idle cells will have to be sent, therefore reducing throughput and increasing latency by increasing the Virtual Output Queue fill level. A reasonable clock frequency for the logic within the switch element and the queue manager is 200MHz today; this is achievable without significant problems in current Application-Specific Integration Circuit (ASIC) technology. Since a 200MHz clock translates into a 5 ns cycle time, the Request/Grant protocol, including the determination of finding a path, must be finalized within an integer number of cycles. Adding round-trip delays adds wasted cycles. A 1m distance between the queue manager and the switch element (or 50 cm of trace plus multiple sets of connectors) adds 6.6 ns round-trip delay, and if the handshake protocol involves the ingress-side queue manager, the switch element, and the egress-side queue manager, then the additional delay for the round trip amounts to 13.2 ns. So, instead of requiring 5 ns total time to carry out the handshake and determine a path, this now will take either 10 ns (a one-sided handshake) or 15 ns of scheduling time. This is independent of pipelining those requests. Unfortunately, during these times fill-in signaling units or idle cells will have to be sent, and the net throughput is reduced. Additionally, the Virtual Output Queues cannot be emptied, and the fill level will not only not decrease, it will increase. Latency through the switch increases, and throughput drops.

As another side effect, the network processor on the line or processor card will be interrupted more often, since the fill levels of the Virtual Output Queue will more often exceed preset threshold levels. In effect, this will decrease the network processor throughput by reducing the link rate utilization.

ROUND-TRIP DELAY (MINIMUM, MAXIMUM, AVERAGE)

The round-trip delay is an important criterion for the ability of ingress- and egress-side traffic managers to minimize queue depths and therefore latency, as well as to avoid oscillations of queue depths between the ingress-side and the egress-side traffic manager. It is determined by the delay and latency of a datagram's transit time through the switch fabric, and heavily depends on traffic patterns, the distribution of priorities, and on the priority of the datagram. In a cell switch, the round-trip delay is additionally influenced by the Reassembly unit and the time it takes for completion of the packet from the cells. The larger the round-trip delay, the longer it takes for the traffic managers to correctly and accurately signal overload conditions, contention, and queue depths. The important parameters for the minimum, maximum, and average round-trip delay determine the quality of the switch fabric.

CELL DELAY VARIATION (CDV) (MINIMUM, MAXIMUM, AVERAGE)

The Cell Delay Variation (CDV) is an ATM-only parameter. It pertains to ATM only because only ATM has a TDM-like traffic type, CBR (Constant Bit Rate), which is used for voice calls. Cell Delay Variation impacts voice calls and unencoded video data only. MPEG-encoded video streams are practically not impacted by the CDV. However, a large CDV will force a large packet delay variation, and for real-time video transmittals it will result in a larger delay between the data sent and the data displayed; as a consequence, it will require deeper buffers on the receiving side. Large CDV will also require an echo cancellation circuit if the router is used to transport voice calls. A router designer should strive for a low packet delay and delay variation, and therefore typically will demand a low CDV from the switch fabric.

The important parameters for the minimum, maximum, and average CDV determine the quality of the switch fabric.

SCALABILITY

Scalability is of foremost importance during the expected life cycle of a router or a switch. There are two major components to scalability. One is that the switch fabric is scalable, and the other is that the node must be able to make use of scaled-up switch fabric components without requiring a forklift upgrade. In order to be useful over the life of the node, the switch fabric, as the core component, must be scalable by architecture. This means that one generation of a switch fabric must span a wide range of port numbers and port speeds without performance bottlenecks while scaling up and without dramatic overprovisioning of performance during the first part of the cycle—and therefore without the additional cost associated with it. The line cards should be independent of the hardware specifics of the switch fabric and therefore independent of the total capacity. As an example, a node could be designed such that it supports a backplane with 16 slots, initially only supporting 8 line cards with 2.5 Gbit/s (OC-48) line cards, thus supporting a total of 20 Gbit/s full-duplex user traffic. This design might then require migration to a 16-port system with either the same cards, quad OC-48 cards, or 10 Gbit/s (either OC-192 or 10 GbE) ports, thus requiring a total switch fabric capacity of 40–160 Gbit/s full-duplex net user bandwidth. It is more than desirable if the line cards do not have to be changed if they are deployed in the larger system, and it is very helpful in terms of software development time if the 2.5 Gbit/s line cards are software-compatible with the 10 Gbit/s line cards. This would require that all switch fabric components are located on the switch fabric card. It also requires that there are switch fabric configurations that span a range from 20–160 Gbit/s net user bandwidth. Additionally, adding throughput should also add VOQs and scheduling performance, so that bottlenecks in scheduling can be avoided altogether if migrating to larger configurations.

The switch fabric partitioning must be such that an upgrade or update to any component leaves the other components unchanged. This is necessary to avoid forklift upgrades. For example, a higher-speed line card that gets swapped in for a lower-speed older line card should not necessitate a hardware change anywhere else. Likewise, if the switch fabric capacity is increased, no line card should be affected. This requires that all switch fabric components reside on the switch fabric card. A switch fabric component residing on a line card very likely is affected by a capacity upgrade—even if it is just the number of ports or the port speed, and thus the queue depth is increased. A switch fabric upgrade must not necessitate any hardware change to a line card. Also, for reliability reasons, all switch fabric components must be covered by the 1:1 redundancy of the switch fabric card. That again requires these components to be assembled on one physical card. Components that are not on the switch fabric card but somewhere else cannot be included in the 1:1 redundancy, and therefore are not covered by the redundancy architecture. This dramatically decreases the system availability.

Scalability additionally means that the number of switch fabric ports can be increased without impacting the line cards that are already installed. This is especially an issue in installations that will grow linearly over time.

Adding a line card means adding a port. If a switch fabric is capable of supporting N ports, then adding one port renders $N + 1$ ports that the upgraded switch fabric must be able to support. Usually, switch fabrics support 2^n ports, and if $N = 2^n$, then the entire switch fabric must be exchanged for a switch fabric with a larger capacity. While this seems problematic, it is not really an issue. This is easily accomplished, especially in installations with redundant cards. The hot-standby card is placed into the status Maintenance Blocked (MBL), subsequently in Unavailable (UNA), and then removed. The new card with the higher capacity is installed and placed into the status MBL. After it is initialized, it can be put into the status Active (ACT), and the remaining card can be handled in a similar way. After everything is returned to ACT and Standby (STB), the system is extended in its switching capacity. That requires that the switch fabric cards contain all components, including the queue manager. If the queue manager is on the line cards, then the number of VOQs supported remains the same, even after upgrading the switch fabric card; therefore, the additional ports cannot be used since there is no VOQ for them. As a result, the router cannot be upgraded by just exchanging the switch fabric card. Instead, the line cards must be exchanged as well.

Scaling up the switch fabric is not a trivial issue. It certainly is limited by the implementation if the current architecture is kept. Die sizes are expected to be huge. On the upside, however, another architecture that has chosen to go with queues at every crosspoint faces this problem to an even higher extent. With VOQ-based architectures, the memory required basically grows linearly with the number of ports N, and, to a lesser extent, with the port speeds. Queues at every crosspoint imply a growth according to N^2. It appears that the growth of single-stage routers is limited by the size of buffered and queued switch fabrics. There may be a shift in paradigm to deploy multistage switch fabrics in routers.

However, we can determine the degree to which scalability is possible in the current paradigm of VOQ or CVOQ switch fabrics. If we assume we will have to linearly increase single-stage switch fabric capacity, we can see the implications. Let us assume that for the time being a net (payload) 5.12 Tbit/s switch is sufficient. The challenge then is to find out which architecture to use. If VOQ architecture is deployed, then the individual queues impose a big implementation problem. At 10 Gbit/s a 1024 cell VOQ is acceptable to prevent Head-of-Line blocking. It is to be assumed that for a 40 Gbit/s solution, 4,096 cells are required to be stored per VOQ, and that number will have to increase if the network processor or the traffic manager has a proportionally higher latency at these speeds. So for a 40 Gbit/s port, every VOQ must be able to hold 4,096 cells. If the current function split is kept, then the queue manager must hold vastly more and faster memory than any current one. It must hold a total of 4,096 cells per VOQ, multiplied by 64 or 128, depending on the internal datagram size. This is a total of 524,288 internal datagrams, or 37,748,736 bytes if 72 byte internal datagrams are used. It is a total of an impressive 301,989,888-bit (around 300 Mbit) SRAM only for the payload in the VOQ in the queue manager. Not only is this significantly more SRAM than deployed today, it also must run at higher speeds. Additionally, it does not yet incorporate any tags, transient LCIs, and Output Queues; nor does it include the logic for the schedulers.

FIELD UPGRADABILITY

Field upgradability requires architectural provisions that provide some level of redundancy, OAM&P support, and scalability of the switch core.

For carriers and backbone network operators, it is absolutely essential that upgrades and service as well as maintenance to routers and switches can be performed while they are in service. A service disruption is very unwelcome. A forklift upgrade should be only required if the backplane breaks. This mandates that any or all components can be easily changed without any adverse effect on any remaining or new components. A router or switch that was designed to support 160 Gbit/s, but was delivered with 8 line cards of 2.5 Gbit/s each and a corresponding switch fabric card with 20 Gbit/s switching capacity, will experience a need for an upgrade in at most a year after its initial installation. It must be possible in the first step to add 8 line cards of 10 Gbit/s each, upgrade the switch fabric card to a 160 Gbit/s throughput, and later replace the old 2.5 Gbit/s line cards with 10 Gbit/s line cards. This process must not require a complete swap of all components in each of the steps. In turn, this requires that all components of the switch fabric be physically installed on the switch fabric card. If switch fabric components are on the line cards, then any upgrade will necessitate a complete swap of all cards—an expensive and disruptive proposition.

RESOURCE UTILIZATION ON THE NETWORK PROCESSOR

Apparently, the choice of the switch core has a fairly large impact on the resource utilization of the network processor. While a crossbar switch may be the cheapest solution of a switch core for a given router architecture, the total system cost may be higher than with a switch core that consists of a self-routing switch fabric. Obviously, a self-routing switch fabric does not require a scheduler for the switch core itself, and it does not use resources on the network processor for the process of switching (i.e., setting up and tearing down connections in the switch core). Ideally, the network processor only performs tasks for which it is uniquely designed. While any network processor must work in conjunction with the switch fabric, the degree of involvement differs. Traffic shaping and traffic management are network processor tasks, as are the classification of datagrams and the attachment of a Local Connection Identifier. The LCI contains the egress port number and the priority or QoS. Any involvement exceeding these tasks puts an unnecessary burden on the network processor. If the switch fabric requires more actions of the network processor to switch datagrams, then the reduced complexity in the switch core is bought with an increase in complexity of the network processor. This is not a good tradeoff since there are more network processors than switch cores in a router and both devices are in the data path; as such both must work and function at line speed. A cost saving of $10 because of reduced complexity in the switch core will then lead to an increase in the complexity of network processor hardware and software that exceeds $10 per port. At 16 ports, a savings of $10 in the switch core (and $20 if it is a redundant core) will lead to additional expenses of at least $160 for the entire router. Instead of saving money, the system becomes more expensive without any benefit to the user.

COST STRUCTURE

While an advanced switch fabric will provide significant benefits to its user, it will have to justify the associated cost. As a result, it must be as efficient as possible. This means that its architecture must be as good as possible for the intended use, and its implementation must be as good as tool technology and silicon allows. An architecture that ignores those basic truths will not perform as intended. An Input-Buffered Output Queued Switch Fabric—no matter how good the implementation is—will not be able to compete with a CVOQ Switch Fabric because, if the same die size is targeted, the CVOQ Switch Fabric will have better contention resolution and a lower latency as the aforementioned, at the same cost. With the same metrics for both switch cores, the CVOQ Switch Fabric can be made on a smaller die or on an older, cheaper process, and therefore is cheaper. If connection setup and teardown times are irrelevant, then both will lose to a crossbar or crosspoint switch. The application in this context determines the architecture—and the cost of the switch core.

FEASIBILITY

The best implementation of a switch fabric is useless if it is not feasible. First of all, the resulting PCB for the switch fabric card must be manufacturable. Second, it must connect to the rest of the system. This in many cases sounds easier than it really is. Since all line cards are connected to the switch fabric card through the backplane or the midplane, there are a huge number of connections to be made. All of these must go through connectors, and there is where the resulting challenge lies. The currently available connectors for HSSL are limited to about 1200 individual connections or pins. The reason for this is the physical size of the connector on the switch fabric card and on the backplane; more importantly, the size and number of connections is limited by the insertion and extraction force that can be applied to the card. Levers do not solve this problem. It does not do any good if the extraction forces required to extract the card will detach the connector from the backplane— destroying the backplane and therefore requiring a swap of the entire device.

Additionally, the power requirements and the heat generation must be within the limits of the design of the node, and thus require the individual switch fabric components to be rather moderate in their power consumption. It also means that all switch fabric components are on the switch fabric card instead of on the line cards, since their power consumption is already at the limit. Another factor is chip count and pin count per chip. Both parameters determine the minimum board size and number of layers on the PCB. High-pin-count devices with parallel buses require high layer counts, making the PCB more expensive to design, more expensive to manufacture, and much more complicated to test. Again, a solution is to use High Speed Serial Links.

PERFORMANCE EXTENSION

There are many ways to increase the performance and throughput of a switch fabric. One is to scale, staging multiple crosspoint or crossbar switches into a multi-stage array of switches. This can be done in a blocking, partially blocking, or a non-blocking fashion. The goal can be to increase the number of ports, to increase the port rate, or both.

If only the port count of the crosspoint switch in the switch core is increased, then the number of VOQs does not correspond with the number of ports any more, and the entire switch becomes subject to Head-of-Line blocking and contention as well as Head-of-Line blocking again. As a result, a VOQ or CVOQ switch fabric must be extended not only in the switch core, but also in the queue manager's ability to supply sufficient VOQ and output queues.

Another way to scale the performance is to deploy multiple slices of switch fabric planes. This can be done by just parallelizing the crosspoint slices and deploying a single scheduler, or by using a scheduler per slice. Slicing can be performed on a per-bit, per-datagram, or per-timeslot basis. Obviously, the requirements vary significantly for the scheduling performance, and the scheduler may or may not be able to keep up. In practically all cases, the egress-side line card might need to reorder datagrams or bits that have been received out of order.

Some schemes call for load sharing among the slices. However, that has proven to be too compute-intensive without any real benefit, and therefore is rarely deployed.

Multi-stage switching can be achieved through Clos, Benes, or Banyan architectures. These are non-blocking performance extensions of a switch. It can also be hierarchical, which is a partially-blocking performance extension. The scalability of the crosspoint switch thereby may be the easiest part of the scaling of the switch fabric.

BLOCKING MULTI-STAGE SWITCHES

A hierarchical multi-stage switch is comprised of hierarchically connected crosspoint switches (see Figure 9.5). Since not all row/column tuples have a crosspoint that can connect them, a hierarchical multi-stage switch fabric is not blocking-free. It is partially blocking. As a result, it does not make sense to reduce contention, congestion, and Head-of-Line blocking to the maximum extent because it is not blocking-free. Should a multi-stage switch be required, and should the partial blocking of a hierarchical switch be acceptable, then it might make sense to hierarchically connectswitch fabrics instead of crosspoint switches. In that case, the VOQs of the individual switch fabrics would prevent at least some of the contention internally.

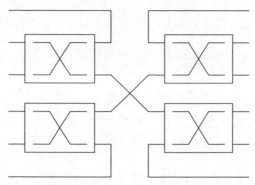

FIGURE 9.5 Hierarchical (blocking) network of crosspoint switches.

Obviously, this cascaded multi-stage switch fabric has a concentration factor of 3:1. That means that any two out of three connections at any time must be made locally, and only one out of the three connection setup requests can be forwarded into the second stage of the switch. Therefore, it is partially blocking.

Non-blocking Multi-Stage Switches

Clos, Benes or Banyan architectures can be cascaded to form a larger crosspoint switch that is blocking-free. Adding or modifying an internal scheduler renders a larger self-routing crosspoint switch. External queue managers with appropriate queue numbers and depths can then be used to provide VOQ and output queues. The queue managers will have to contain schedulers to enqueue and dequeue the data. A Clos network of crosspoint switches would be a very typical blocking-free cascaded architecture (see Figure 9.6). In this example, all connections can be made locally or globally, staying in stage 1 or being forwarded to stage 2 and routed to the appropriate egress port. It is blocking-free.

Bit Slicing

In a performance extension architecture that deploys bit slicing, multiple slices of the crosspoint switch are parallelized (see Figure 9.7). The queue managers send the individual datagrams to all slices on a per-bit basis in a round-robin fashion. This only makes sense with an external scheduler that simultaneously sets up the slices, and only if the slices are microcycle synchronous. The bits must be reassembled on egress, but further synchronization and re-ordering should not be necessary. Failure of any of the bit slices leads to a total loss of the datagram, since reassembly becomes impossible. Loss of synchronization between the switch fabric planes also leads to a total loss of datagrams.

Cell Slicing

In a performance extension architecture that deploys cell slicing, multiple slices of the crosspoint switch are parallelized (see Figure 9.8). The (external) queue managers send

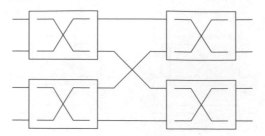

FIGURE 9.6 Non-blocking multi-stage crosspoint switches.

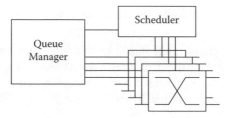

FIGURE 9.7 Bit-sliced array of crosspoint switches.

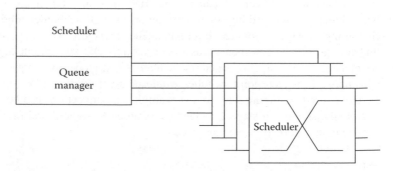

FIGURE 9.8 Cell-sliced array of crosspoint switches.

the individual datagrams to all slices on a per-cell basis in a round-robin fashion. The slices do not have to be synchronized at all since Reassembly and re-ordering on egress is necessary in any case. As a result, all slices can be self-routing switch fabrics or externally controlled switches. If self-routing non-buffered, non-queued or CVOQ switch fabrics are deployed, contention can be minimized. For CVOQ switch-fabric-based slice architectures, the CDV within and among the slices should be bound to a certain maximum to ensure that correct and timely packet Reassembly is possible.

LOAD SHARING ACROSS N PLANES

In a load-sharing application of N switch fabric planes (see Figure 9.9), the queue managers send the datagrams to the switch fabric plane with the lowest load. While it

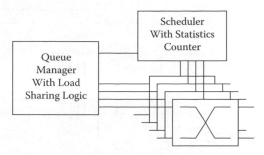

FIGURE 9.9 Load sharing across crosspoint switches.

increases total switch fabric throughput, it demands a significantly higher scheduling performance from the queue managers; it also relies on a protocol between the switch fabric planes and the queue managers to determine the load and the VOQ status, output queue status, and in some cases the status of the egress link. While the process of striping is pretty simple, reassembling the fragments on egress and reordering the fragments to make cells or packets from them effectively reduces the advantages of the performance upgrade. Either a massive reordering logic with huge memory must be installed in egress, or all planes must be synchronized on a per-bit basis. This would prohibit any reasonable scheduling inside the switch fabric planes and therefore appears to be limited in its deployment. Additionally, in case of failure of one plane, either all planes are out of operation or the slicing and recombination/reordering scheme must be updated in real time through a non-functioning switch fabric. This promises to be a very complex task, especially in carrier class environments where five nines (99.999% uptime) or six nines (99.9999% uptime) are required. A non-redundant load-sharing architecture by itself is not fault-tolerant at all. Any fault will either take all $N + 1$ planes out of operation (which then is not a redundancy scheme at all), or the propagation of the fault information is insecure, making the switchover or the exclusion of the defective plane a matter of chance instead of a deterministic process. Even if that is not considered, the reordering logic still must be on the line cards, and that makes the line cards more complex and therefore more expensive—and they dissipate more power. This is the opposite of the desired goal.

MASTER/SLAVE OR "PERFORMANCE EXTENSION ARCHITECTURES"

Master/slave architectures (see Figure 9.10) are defined by a multiplicity of similar or identical switch fabric planes in which one of the planes is designated to be the master and the remaining planes act as slaves. The master plane thereby receives the control path information and sets up the connections within all planes—the master plane and the slave planes. Master/slave architectures have a higher scheduling latency since the master scheduler must communicate with the slave switch planes to make sure a path is available. Because this is off-chip communication, it is much slower than internal or logically distributed scheduling and switching. Additionally, master/slave architectures prohibit a graceful degradation in case the master is defective. Master/slave architectures, by design, are not fault-tolerant. The

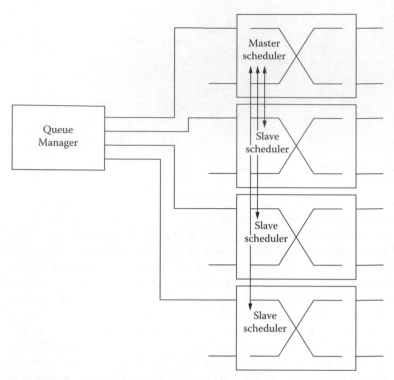

FIGURE 9.10 Master/Slave architecture of crosspoint switches.

master is the single point of failure if there is no hot-standby plane with a similar scheduler. In some cases, a slave is set up as a backup master in case the master fails. They require a 1:1 redundancy, contrary to popular belief and claims. The master poses a serious single point of failure without additional measures. Sometimes master/slave architectures are called Performance Extension Architectures. QoS cannot be guaranteed, and CDV typically is unpredictable.

ALTERNATIVE SOLUTIONS

Many customers desire a "pay as you grow" solution for routers and switches. This is possible to a certain degree, but scaling a router will make the initial scaled-down expandable version more expensive than a fixed-configuration, small, non-scalable version. Even history has shown that providing room for growth was not the way to go; for example, Compaq had EISA-based servers for a while that allowed the CPU to be swapped out and upgraded, and they claimed it solved the problem of server obsolescence after three years. It did not solve the problem for a variety of reasons: The NorthBridge, the memory, and all I/O, including storage, remained the same throughout the life of the server, and therefore limited the performance from day one. Additionally, since the most expensive part was the CPU anyway, exchanging the CPU—that had to be mounted on a daughterboard to achieve the goal of being user-replaceable—was about as expensive as buying a new mainboard or even buying a new server. It added cost to the system without any real benefit because the performance never really scaled with the CPU. The same is true for routers; if all provisions are made to accommodate a scalable, but replaceable switch fabric card, then the systems' initial cost is rather high, especially if seen on a per-port cost. Additionally, swapping out the switch fabric for a higher capacity version makes the old one obsolete. Therefore, the only viable solution is to provide a backplane and a full-scale switch fabric card that truly scales, and underpopulate only the line card slots. In this way, the initial investment is as low as possible, but nevertheless retains all options to upgrade the switch fabric cards while adding line cards. This only works if the incremental cost of the line cards is as low as possible, since the line cards are sold in higher volumes than the switch fabric cards. As a result, many people have thought about truly scalable routers that may or may not be based on a switch fabric. The most famous attempts were the "pay-as-you-grow solution" and the "switchless switch." The "pay-as-you-grow solution" asked for a switch fabric card that could be extended in its switching capacity. We will see in the following figures why that did not work in real life. The second proposal was the "switchless switch." We will come back to the "switchless switch" after the figures of the "pay-as-you-grow solution."

"PAY-AS-YOU-GROW SOLUTIONS"

The "pay-as-you-grow" solution requires a switch fabric card that has a basic switch module and a variety of dummy modules that connect the traces that otherwise would not be connected. These dummy modules must be replaced by switch fabric modules if the switch fabric card must be upgraded. For each upgrade, the dummy modules must be changed because the interconnecting schemes change with more switch fabric modules installed. The number of interconnections on the switch fabric card is significant, and the number of connections on the dummy modules is as significant as the connections on the module. Additionally, dummy modules must be changed together with switch fabric modules on the card. The total complexity and the number of different cards that must be held to initially install and then upgrade the switch fabric cards is large, and therefore directly counters the intended and desired cost savings. As a result, "pay-as-you-grow" solutions do not save money—neither upon initial installation, nor on the growth path (see Figures 9.11 to 9.18).

THE "SWITCHLESS SWITCH"

The "Switchless Switch" has been an object of desire for a long time. Designers want the throughput and ease of use of a CVOQ switch fabric, without the expense of it. As a result, more than once a full mesh has been proposed as an interconnecting medium. While it does not require any logic in the "switch" core, it significantly adds to the complexity of the line cards. Additionally, it requires an extraordinarily high number of interconnecting traces on the backplane or the midplane. In a router

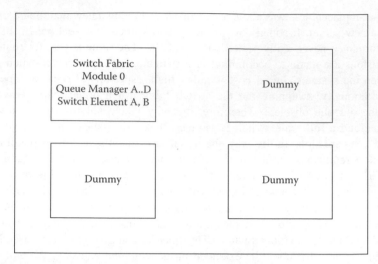

FIGURE 9.11 "Pay-as-you-grow" switch fabric card, quarter populated.

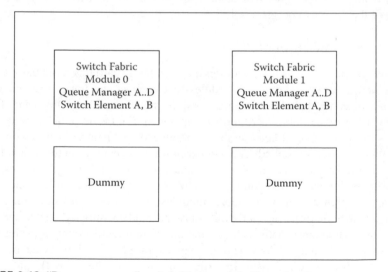

FIGURE 9.12 "Pay-as-you-grow" switch fabric card, half populated.

with n line cards, each line card must provide $(n-1)$ internal egress and $(n-1)$ internal ingress ports to support the full mesh.

It requires $\frac{1}{2} * n * (n-1) = \frac{1}{2} * (n^2-n)$ connections (and therefore the problem is of the order of magnitude of n^2) instead of n in a star configuration deployed in architectures with switch fabrics. As an example, a very typical 16-port router with a switch fabric requires 16 link bundles to go into and come from the switch fabric card. Each line card has exactly one link bundle going in and from the card. In a full mesh configuration, the number of link bundles going into and coming from each line card is 15. However, on the backplane (or midplane), 120 link bundles

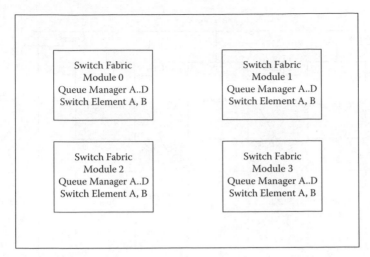

FIGURE 9.13 "Pay-as-you-grow" switch fabric card, fully populated.

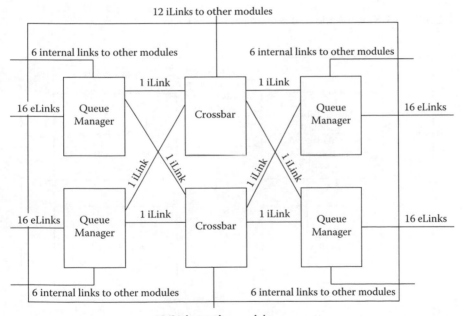

FIGURE 9.14 Switch fabric card connections—link view.

must be routed and connected. For a 10 Gbit/s line card, a link bundle typically consists of 3.125 Gbit/s differential pairs with 8B/10B encoding and two signal ground traces for each differential pair, thus requiring 16 traces. 120 link bundles therefore require 1,920 traces. These 1,920 traces do not even contain the number

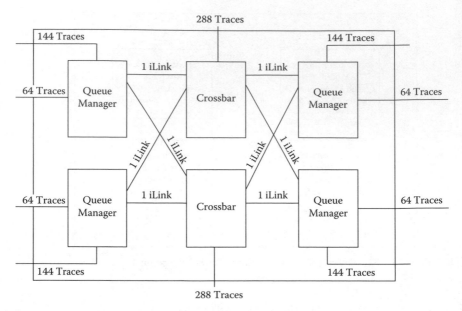

FIGURE 9.15 Switch fabric card connections—trace count.

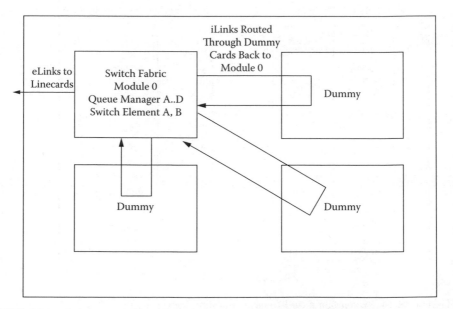

FIGURE 9.16 "Pay-as-you-grow" switch fabric card, quarter populated, connections.

of traces required for OAM&P connectivity, nor do they include any power and ground connections.

The "Switch Core," as a result, does not contain any logic. More importantly, it is entirely passive. However, it is not easy to manufacture, and it will require

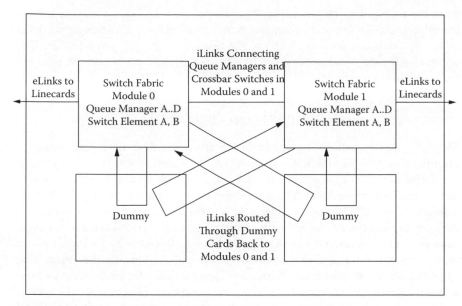

FIGURE 9.17 "Pay-as-you-grow" switch fabric card, half populated, connections.

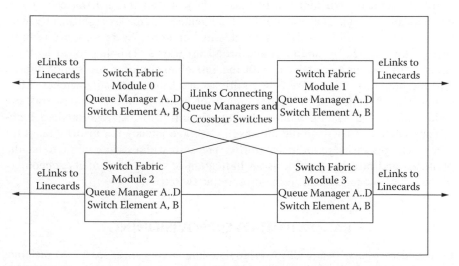

FIGURE 9.18 "Pay-as-you-grow" switch fabric card, fully populated, connections.

a very large number of layers in the PCB for the backplane or the midplane. Additionally, all of the challenges a switch fabric would face are shifted onto the line cards. With each line card having to handle $(n-1)$ internal ingress and egress ports, it must forward datagrams to the appropriate port; more importantly, it will have to handle congestion, contention, and Head-of-Line blocking on its internal ingress port. As a result, each line card will require the equivalent of a queue manager and a traffic manager to handle excess internal traffic offered from the multiplicity of line cards. That makes each and every line card significantly more complex and expensive than a line card for a switch-fabric-based router. In summary, the complexity of the line card—the part produced in higher volumes—for the "Switchless Switch" significantly increases and the total system cost drastically increases.

MULTICAST AND BROADCAST

Multicast and broadcast are methods to distribute data from one source to multiple destinations simultaneously. ATM, IPv4, and IPv6 all have provisions for multicast and for broadcast. By definition, multicast and broadcast are $1:n$ traffic distribution methods. The difference is that multicast is targeted towards a select subset of destination addresses, whereas broadcast goes to all destinations in the subnet or the network. However, neither multicast nor broadcast have been used as intended to the extent that was foreseen by its inventors; the reason was that buses and crosspoint switches could not handle the implications from multicast and broadcast. With early designs, the blocking nature of buses and the inability of external schedulers of crosspoint switches to schedule the connections made the deployment of multicast and broadcast not feasible. While a crosspoint switch is naturally blocking-free, it is not free of contention. This is true for unicast, but more so the case when one datagram from one source has multiple targets. The probability of contention increases significantly with the number of simultaneously targeted destination ports. As a result, even switch fabrics are affected by multicast and broadcast traffic. This has to do with the egress port contention, and not with the implementation of the switch fabric. Full meshes suffer even more from multicast and broadcast than switches. Most switch fabric vendors state the percentage of multicast and broadcast traffic a switch fabric can sustain without losing its ability to losslessly switch high-priority unicast traffic, and the increase in average latency per priority class for unicast and broadcast or multicast traffic. The ability of a switch fabric to handle multicast and broadcast traffic is an indication of the quality of its schedulers in the queue managers and in the crosspoint switch.

BANDWIDTH OVERPROVISIONING

In a modern router, bandwidth overprovisioning occurs in at least two different locations. Both pertain to the switch fabric card, and both help improve parameters such as Head-of-Line blocking prevention and link rate utilization.

Head-of-Line blocking can be prevented with a combination of Virtual Output Queues—which are physically located in the queue managers on the ingress side—and internal bandwidth overprovisioning between the queue managers and the crossbar switch. An overspeed of a factor of two in conjunction with Virtual Output Queuing will decrease the probability of Head-of-Line blocking to nearly zero with Gauss- or Poisson-distributed (destination port and priority tuple) traffic. Bandwidth overprovisioning between the switch fabric queue manager and the switch fabric crossbar switch will require either a larger number of aggregated link bundles or higher rate HSSL between those units. Output Queuing can reduce contention on the egress link of the switch fabric into the egress-side traffic manager or network processor, and makes the most sense if combined with an internal overspeed between the switch fabric and the egress-side queue manager.

In addition, very often the links between the port card and the switch fabric support a higher net payload data rate than the egress port link. This is useful in conjunction with traffic managers, and helps to achieve a better saturation of the links into the switch fabric (inbound traffic) and into the egress link (outbound traffic). The better the traffic manager is able to load the switch fabric's queue manager, and therefore avoid situations in which no traffic is offered and bandwidth through the switch fabric is wasted, the better the link saturation is on the outbound links.

TRAFFIC MANAGER FUNCTIONS VERSUS QUEUE MANAGER FUNCTIONS

Many router designers misunderstand the purpose of the traffic manager on the line card or processor card, and the purpose of the queue manager of the switch fabric chipset on the switch fabric card. They fulfill fundamentally different functions.

The traffic manager typically shapes into the egress link or channel. Its outbound channel capacity is essentially constant. Therefore, its utilization largely depends on the offered load. To a much smaller degree, it depends on the ability of the receiving side to accept traffic. As a result, the utilization of the channel is at its maximum if the traffic manager always has some enqueued low-priority, non-timing-critical data— ready to be dequeued and sent whenever there is not enough offered load from the switch fabric to fill the transport capacity of the channel. As a result, the traffic manager that shapes into the egress link will achieve a link utilization of about 1 (or 100%).

The queue manager has a fundamentally different purpose. The channel capacity into the crossbar switch of the switch fabric is highly variable. Since a crossbar switch is blocking-free, the queue manager only must handle preventing contention and Head-of-Line blocking. However, since there are multiple different priorities of traffic, it must handle the offered load according to its priority. The queue manager must ensure that data is scheduled into the crosspoint or crossbar switch with the lowest possible latency, and it, in conjunction with the scheduler in the crossbar switch, must make sure that a path is available.

SWITCH FABRIC QUEUE MANAGER

In order to prevent contention and Head-of-Line blocking, the interconnecting entity—mostly a blocking-free and lossless switch fabric—will have to provide internal links that are providing higher bandwidth between the switch fabric and the line card. The datagram arrival rate at the ingress switch fabric port cannot exceed 1 significantly, even with an ingress-side traffic manager. The datagram departure rate at the egress switch fabric port cannot significantly exceed 1 either. However, the offered traffic towards the egress switch fabric's port may very well dramatically exceed 1. The switch fabric can be seen as a multiplexer with an internal speedup significantly larger than 1. At any given point in time one or more switch fabric ingress ports may offer traffic towards any switch fabric egress port. Each switch fabric egress port can be the destination of two or more internal datagrams arriving simultaneously at switch fabric ingress ports, whereas other switch fabric egress ports may not experience any traffic at all. As a result, the offered load towards any egress port on the switch fabric may exceed 1. Since the switch fabric and its internal interconnects have significantly overprovisioned internal bandwidth ("overspeed"), the datagram arrival rate and therefore the offered load from the switch fabric egress port towards the line card will from time to time significantly exceed 1 Erlang.

Combined Virtual Output Queuing requires Virtual Output Queues on ingress and egress side Output Queues. None of these must be confused with traffic management functions. The purpose of the switch fabric queuing is to prevent Head-of-Line blocking and to reduce contention, whereas traffic management shapes traffic on the line cards. This can occur into either the switch fabric or the egress port. Traffic management is intended to reduce peak loads of offered traffic by shaping the ingress traffic such that high-priority traffic is forwarded as soon as channel capacity is available into the switch fabric or egress port, and lower-priority traffic is enqueued if the channel capacity is insufficient. Dequeuing of datagrams will occur as soon as excess channel capacity is available.

DETERMINISTIC BEHAVIOR

There is a perception among some router designers that buses are deterministic and switch fabrics are not. Both of these assumptions are incorrect. A bus is a shared resource, and therefore must be shared among all resources. More importantly, blocking and contention cannot be prevented. While a switch fabric is blocking-free, but not contention or Head-of-Line blocking-free, a bus as a shared resource is not blocking-free, contention free, congestion free, nor is it free of Head-of-Line blocking. As a result, neither bandwidth nor latency can be guaranteed through a bus. Not even the response time can be guaranteed because the arbiters are a centralized entity; therefore, scheduling is subject to the round-trip delay between the requesters and the arbiter. The scheduling of the bus depends on the total traffic offered, but more importantly, each transmittal request depends on the total traffic offered, and not only on the traffic to or from one line card. Since the bus is not a blocking-free resource, only one transaction can occur at any given time. More importantly, in a switch fabric that supports multiple priorities associated with different datagrams, scheduling high-priority datagrams ahead of lower-priority datagrams, when compared to the shared bus with a centralized arbiter, renders a better deterministic behavior of the switch fabric.

SWITCH FABRIC I/O

Switch fabric I/O is an important issue since the switch fabric is a core component in a router and the switch fabric determines the internal throughput of the router.

As a result, the I/O technology deployed in the switch fabric determines its through-put, its maximum size, and its reliability in terms of its Bit Error Rate (BER). The higher the bit rate per link, the more datagrams per time unit can be fed through the connector into the switch fabric card. Consequently, an increase in bit rate per link will result in an increase in the maximum switching capacity that the switch fabric card supports. Coding and Error Correction are used to increase the reliability of the transmission line. Current HSSLs are all self-clocked with Clock and Data Recovery (CDR) units in the transceivers, eliminating problems with clock skew and a host of other issues that have plagued wide buses.

Despite the fact that most advanced routers deploy HSSLs, the link rate and the connector limit the switching capacity of the switch fabric card. Even if HSSLs are deployed, the line card to the switch fabric card trace count can easily be determined. In most HSSLs or bundles thereof, the links are clocked at 2.5 Gbit/s, and there is a known type of line encoding such as 8B/10B. This overhead can be determined, and the net payload data rate can be obtained by multiplying the line rate with the line efficiency. As an example, let us consider a 320 Gbit/s net user data rate system. With the queue manager on the switch card, the designer will need 8 links of 2.5 Gbit/s baud rate per line card towards the switch fabric card, providing a 20 Gbit/s aggregate baud rate. This translates into 16 Gbit/s data rate, of which 75% are user data. This leaves us with 12 Gbit/s net user payload data rate (i.e., a 20% port overspeed), and it requires 8 links per line card. Consequently, the port overspeed would be 50% with a 3.125 Gbit/s link under the same circumstances. Since four traces make up a link, and 32 line cards are required to connect to the switch fabric card, there is a total of 4 * 8 * 32 traces (1,024 traces) into the switch fabric card. This does not change if the designer deploys redundant switch fabric cards, because then each of the planes connects via the aforementioned 1,024 traces. A 1,024-pin connector can be implemented. The internal overspeed of the switch fabric is not impacted by this. The situation is different, however, for the architecture with the queue manager on the line card. Here we must take the internal switch fabric overspeed into consideration, not the port overspeed. If we combine port and internal overspeed to achieve the same internal overspeed in case the queue manager is on the line card, then we require not 8 links at 2.5 Gbit/s per nominal 10-Gbit/s link, but 16. Now the requirement is to route 16 * 4 * 32 traces across the backplane. These 2,048 high-speed signals somehow must get into the switch fabric card, and that is where the problem lies; it requires a 2,048-pin connector to connect the backplane to the switch fabric card. This does not even take into consideration that there will be power connectors and lower speed signals for OAM&P routing. Such a connector is not possible right now, and it does not appear that a connector for 2,048 individual connections can be made that fulfills any of the mechanical require-ments of the insertion and extraction forces. Even optical interconnects do not provide enough throughput, and it would not solve the problem of bringing these signals into the individual Integrated Circuits of the switch fabric chipset.

In all modular designs the backplane and the midplane are currently the limiting elements of growth of throughput. This restriction results from the fact that electrical connections are more throughput-limited than optical connections, and that the insertion and extraction forces have too-limited to manageable values. In modern

HSSLs one link can carry symbol rates from 3.125 Gbit/s to around 11 Gbit/s. However, it must be seen as four pins on a connector per direction: D+, D-, and two GND to shield. As a result, a bidirectional link with 3.125-Gbit/s symbol rate requires eight pins. The 3.125-Gbit/s symbol rate most often translates into 2.5 Gbit/s net data transfer rate because of the 8B/10B coding used. A modern router will always provide some internal speedup between the line cards and the switch fabric card, and therefore will require a higher data rate provided internally than is available for the external links. Typically, a factor of two is used to avoid Head-of-Line blocking, and therefore 16 pins for a total of 5 Gbit/s internal capacity between the line card and the switch fabric card will be required for a net 2.5 Gbit/s external link. With a preset maximum of 1,000 pins per connector into the switch fabric card, this will provide enough capacity for 62.5 ports—not counting any other signals that might be required. Considering power and GND and other signals required, a reasonable number is 32 external ports of 2.5 Gbit/s net data transfer rate. In other words, with current HSSLs a router can be built with 32 OC-48 ports or with 8 OC-192 ports. If higher numbers of ports or higher data rates per port are desired, the number of pins on the connector into the switch fabric card or the symbol rate on the HSSL must be increased. Implementing 16 ports of OC-192 with current HSSLs requires segmenting the connector into the switch fabric card into the signal portion with more than 1000 signal pins, and using a second connector for power and GND.

Another solution is to use HSSLs with a higher symbol rate. These have become available very recently, and provide 11 Gbit/s of symbol rate using 64b/66b coding. As a result, a bidirectional link carrying 20 Gbit/s for an external line rate of 10 Gbit/s net data rate requires 16 pins under the same circumstances as in the above example. This would provide enough internal throughput to support roughly 32 OC-192 (or 10 GbE) line cards or 8 OC-768 line cards.

SOFTWARE FUNCTION SET IN LOCAL SWITCH FABRIC CONTROL

One could—and should—argue that on a switch fabric card, an additional local control processor not in the data path does not have a significant cost impact and it provides additional capabilities for local control and bootstrapping, control interface hardware to the OAM&P card, and means for statistic traffic monitoring. While the cost and surface area are immediately clear, the software effort is not. Typically, the software development effort for these local control processors is in the order of hardware development cost of the digital portion of the board—but this is not the whole story. If there is some local control on the switch fabric card, then the centralized OAM&P card can be divested of a few functions that do not need to be processed device-globally. As a result, the OAM&P card's software can be reduced in complexity and improved in availability and robustness without impact on functionality.

Software on the switch fabric card is not in the data path. As a result, there is no requirement for real-time response of any microprocessor or microcontroller software on the switch fabric card. However, this does not mean that the robustness and the fault tolerance of the microprocessor or microcontroller software on the switch fabric card are not important. The opposite is true. First of all, the software must setup and configure the hardware in its functions and features. The software also must perform statistic traffic data collection for traffic monitoring and traffic engineering.

It is of no use if the hardware by itself is crash-proof but the software running on it frequently crashes or must reboot due to software faults. That means that the software must be capable of surviving some hardware errors. Garbage collection must be flawless. In case something goes wrong anyway—like spurious hardware faults or nested Interrupt Service Routines that terminate because of run-time issues—a planned shutdown and restart (recovery) must be implemented. However, rolling recoveries must be avoided.

Functions that are typically executed on a local control processor on the switch fabric card are bootstrapping, setup and configuration of the switch fabric chip set, statistic traffic monitoring, error statistics and heuristics, and sometimes even more complex traffic analysis of data provided by the switch fabric chip set, such as: number of routed packets or cells over a preset period of time per priority class; dropped cells within that period of time per priority class; the minimum, maximum or average number of queued cells or packets in the switch fabric per priority class; the minimum, maximum and average duration of queue or switch fabric traversal time or even scheduler information per priority class. BERs and cell or packet error rates can be provided by the switch fabric chip set as well and in that case will be collected by the local control processor for completion of the statistic traffic monitoring to be forwarded to the centralized OAM&P card. With information like that, it is fairly simple to prove QoS fulfillment for true IPv6 compliance.

The software in the local switch fabric controller must be able to carry out the following most basic tasks:

- Setup
- Initialization
- Statistic Traffic Data Collection and summarizing the individual results from Queue Manager and Switch Elements
- Local error and status collection
- Local event handling within administrative statuses
- Status supervision and assignment (ACT, STB, DEF, UNA, and MBL)
- Communication with and reporting into OAM&P card
- Watchdog timer operation for local control

Therefore, it requires that the local controller communicates with the OAM&P card and can parse its messages and react upon them. It will also require a watchdog timer in order to make sure that the software can automatically be restarted in case of a software crash. There also should be a set of setup commands or APIs that enable the user of the software to set up the switch fabric in a few predefined robust default and standard configurations. One of these could be 16-port 40-Gbit/s configuration with strict priority in the queue manager and round robin in the crosspoint switch, and additional commands that could be used to change the setup in a finer granularity. The setup also should include an error notification system and basic functions to communicate with the OAM&P card.

The interface between the line card and the switch fabric card is restricted to the combined data path and control path for the payload and the header or LCI. There are no commands sent from line cards to the switch card. In other words, the switch card does not terminate any traffic. Commands into the switch card originate only from the OAM&P card. Statuses are sent from the switch card into the OAM&P card. If necessary, the OAM&P card sends statuses or commands to the line cards, but it does it out of band. The switch fabric card does not act as a host; it is the client.

There is no API required in the line cards to handle switch fabric card messages, statuses, or commands directly. The OAM&P card or the OAM&P task on one line card will handle all this and forward it to the switch fabric card if necessary, when appropriate.

The local switch fabric controller software must support all OAM&P functions. This is necessary to enable Traffic Management and Engineering. The OAM&P functions are run and carried out by the OAM&P card; they consist of functions that collect statistic data, functions that enable the switch to operate, and functions that handle errors and exceptions. OAM&P functions are crucial in a router or a switch. These functions typically are implemented in a centralized OAM&P card, but not on the switch fabric card.

In addition to these, the local control processor and its software must be able to set and recognize and react accordingly to the following statuses:

- ACT (active)
- STB (standby)
- DEF (defective)
- UNA (unavailable)
- MBL (maintenance blocked)

All status transitions are received from the OAM&P card, and the switch fabric status can be sent back. However, the switch fabric controller is not supposed to change the administrative status by itself. It is only allowed to receive set status commands, verify them, and place the switch fabric plane into the appropriate status. It also should diagnose the switch fabric chips and send notifications to the OAM&P card. ACT is the active state, and the OAM&P card can send a set status command from ACT to any other status. There is only one ACT switch fabric card; the others are in any of the other statuses. A transition from UNA or DEF to ACT is not allowed. DEF or UNA must be transitioned to MBL first, then to STB. Transitioning one switch fabric plane from STB to ACT means that the other plane that was previously in ACT will automatically be placed into STB unless it goes into DEF. It is required for the switch fabric controller to hold its status in local memory, compare incoming status transition commands with the allowed set of transitions, implement it, and send a status update message back to the OAM&P card.

There must be APIs or function calls that easily handle these statuses and the transitions. Additionally, there must be a signal for the switch fabric controller to the OAM&P card, used for notifications in case of locally reported errors. This can be done using messages or optionally by asserting an IRQ towards the OAM&P card.

In addition to these functions, there should be functions or APIs to reset individual links, queue managers, and the switch elements.

Finally, the watchdog timer should supervise CPU activity. If the CPU does not reset the timer within a predefined period of time, then the timer expires and resets the CPU via its RESET signal so that the software can reboot. It may make sense that the software can be reset via command, too.

CONCLUSION

Switch fabric applications differ drastically based on their intended use. In some applications, changes of the setup are rare and switching can be considered static or quasi-static. Under those circumstances, a crosspoint switch or an optical cross-connect can be the perfect solution.

Switch fabric applications requiring high bandwidths while maintaining low latency, QoS awareness, and low cell losses are typically found in TelCo-grade datacom environments. Data communications over these devices have requirements that match or exceed NEBS standards in terms of performance, throughput, reliability, and system availability, upgradeability, and scalability. Throughput must not be compromised under any circumstances; Quality of Service must be maintained in order to fulfill the Service Level Agreements.

These requirements are reflected in current switch fabric architectures, which are at the heart of the current and future generations of routers and ATM switches. The predominant switch fabric architecture that has emerged out of this is the Combined Virtually Output Queued (CVOQ) switch fabric. It is an Output Queued (OQ) and Virtually Output Queued (VOQ) cell-based switch fabric with queue managers, crossbar switches and other components that provide the memory and the scheduling or arbitration, together with Serializer/Deserializer (SerDes) components. Some implementations of this architecture combine multiple functions into one single chip; others separate these physically or logically. Mostly, the queue managers incorporate the schedulers and the memory required for the OQ and the VOQ functions. In some instances, the queue managers incorporate High Speed Serial Links (HSSLs) to connect to other line card components and the crossbar switches.

In the above-mentioned CVOQ switch fabrics, the queue managers can be placed either on the line cards or on the switch fabric cards. Both placements seem to have good reasons. However, one placement makes more sense than the other.

We have shown that in router architectures with VOQ, OQ or CVOQ architectures work better with switches that have the queue managers and the crossbar switches combined on the switch fabric card. They offer better throughput, lower latency, higher levels of link rate utilization, and distribute power dissipation better and more evenly over the entire device. They therefore avoid producing hot spots. This also contributes to the possibility of incorporating line cards with combined throughput rates that are lower than the native queue manager line rate. Putting more logic and memory onto the switch fabric card significantly reduces the amount of logic and buffers on the line cards. Since there are more line cards than switch fabric cards, this additional effort on the switch fabric card will be overcompensated by the reduced amount of logic and memory on the line cards.

10 Operation, Administration, Maintenance and Provisioning

OVERVIEW

Telecommunication networks have historically relied heavily on the administration of user features and their access to the network. Data networks completely negated this approach and made a wide variety of services available to users, including self-administration of features and access. As has become evident over the last few years, complete self-administration of user features, user access, and administrative rights without authentication and without any accountability is not working properly anymore. Spam, fake sender identification, and intrusion into networks and data centers have been made possible by a variety of factors, including the lack of at least a basic level of Operation, Administration, Maintenance, and Provisioning (OAM&P) for network access rights. Additionally, traffic engineering in data networks has largely been unsuccessful due to a lack of insight into the networks' load status and the nodes' statuses. As a result, in future advanced networks—Internet2 and other, smarter integrated networks—OAM&P will be of paramount importance in advanced router designs. It is a crucial hardware, firmware, and software building block in current and future networks and all nodes thereof.

In the Introduction chapter, we derived requirements for advanced routers from the desires of customers, carriers, and ISPs, and most of the requirements have a direct impact on the OAM&P entity. Future routers will have to be able to meet the following requirements:

- Losslessly switch and route datagrams
- Be able to perform Segmentation And Reassembly (SAR)
- Perform policing at line speed for thousands of virtual connections simultaneously and thereby enforce and support SLAs
- Perform Traffic Management by queuing and buffering traffic according to SLAs, and drop excess traffic
- Support Traffic Engineering by means of collecting Statistic Traffic Data
- Support a mixture of hierarchical and mesh interconnect infrastructure in terms of data traffic
- Support an overlay network for metadata and signaling data
- Enable traffic rerouting at the edge and provide redundant fail-safe systems towards the core
- Communicate securely within the components of the router

- Communicate securely between the OAM&P card and a Billing Center
- Communicate securely between the OAM&P card and a PKI Center for authentication
- Communicate securely between the OAM&P card and a Network Management Center (NMC)
- Communicate with the PSTN infrastructure

The list of requirements that directly determines which functions of the OAM&P entity those affect is as follows. The OAM&P entity has to

- Support Traffic Engineering by means of collecting Statistic Traffic Data
- Communicate securely within the components of the router
- Communicate securely between the OAM&P card and a Billing Center
- Communicate securely between the OAM&P card and a PKI Center for authentication
- Communicate securely between the OAM&P card and a Network Management Center (NMC)
- Communicate with the PSTN infrastructure

With this is mind, we can define an OAM&P entity and what functions it will have to support.

DEFINITION OF OAM&P

The main purpose of OAM&P is operation support, administration support, maintenance support, and provisioning support.

Operation support functions include software routines to allow for remote management of the system, including supervisor access through the local ports or through the network using FTP, TFTP, RMON, or through a variety of proprietary protocols. They are also used for setting user administrator access rights through Access Control Lists (ACLs), and for blacklist and whitelist management. They are used to update firewall functions as well as existing and new blacklists and whitelists. Traffic management and engineering is performed through operation support functions. Operation functions for statistic traffic data collection, including billing information, enable the OAM&P entity to download the collected data from the individual modules and upload them into the NMC. The administrative functions include software to setup all components, building blocks, or modules and to delete, modify or copy routing tables and policies, especially in the network processors on the line cards. They are used for determining, setting, or changing the operational statuses of all components, building blocks, or modules for purposes of scheduled or emergency maintenance. Maintenance functions help determine and configure each component and building block into one of the status active (ACT), standby (STB), maintenance blocked (MBL), defective (DEF), or unavailable (UNA), and to perform a routine check by providing switchover commands from active to hot standby and vice versa for redundant components or modules—running routine checks on all of the nodes' cards internally and locally—and to report the health status of the device and all components. Administrative functions keep database entries of the current statuses and submit these database entries to the NMC. They are also used for receiving network-global and node-specific commands from the NMC, and for executing these commands locally. In a redundant subsystem, they are used for initiating regular timed or emergency switchovers of redundant components or modules. Maintenance support is required for error reporting and recovery from errors. The maintenance functions are also used to reboot subsystems or modules, shut them down, or restart them. It also covers functions to recover from software failures, crashes, and shutdowns. After a reboot or restart, they are used to initialize and populate routing tables and policies. Maintenance functions are available to scrub memory and purge collected data after a software crash in the module. The provisioning support is intended to provision physical and logical ports and interfaces to customers or between nodes, including their specific configuration. Provisioning makes services available or removes accessibility, mostly involving user signaling after services have been configured through administrative commands.

These functions work in conjunction with all major components and building blocks of the router or switch. The OAM&P card polls the status of these periodically—typically in periodic intervals of 15 seconds to 15 minutes—and then processes them locally so that they reflect the device status, compared to the module's status. It then sends them to the NMC, where the data is used to provide input for further traffic management and traffic engineering purposes. In order to do so, all data from all components is collected into a Management Information Base (MIB), which in turn is made available to the NMC. The NMC then uses the routers' MIB data for a network-global MIB. OAM&P, MIBs, and network management in general are defined in a variety of Internet Engineering Task Force (IETF) and ITU-T standards. Simple Network Management Protocol versions 2 and 3 (SNMPv2 and SNMPv3), as well as Q_3 together with the appropriate MIBs are described in IETF RFC2570 and 1905 and others as well as in CCITT/ITU-T Q.811/812 (International Telecommunications Union–Telephony Section) together with Common Management Information Protocol/Common Management Information Service (CMIP/CMIS) in CCITT/ITU-T Q.710/711/712.

The administrator in the NMC uses the OAM&P functions to logically make resources available, to verify SLA compliance, and to ensure network security.

FUNCTIONS OF THE OAM&P ENTITY

The OAM&P entity is responsible for monitoring the health of all modules within the router itself, and it sets and holds the current status information of the router. It may or may not be in a redundant configuration itself. It determines and sets the operational status (ACT, STB, MBL, DEF, and UNA) of the routers' components. The OAM&P entity polls the components and then assigns their statuses based on the information it has gathered. At first, it appears counterintuitive that device-global local supervision is required and that the modules or components cannot determine their own operational status. It would appear more logical that each and every component that can have an operational status should determine its operational status, and assign it. However, this is not useful, since the modules do not have the device-global system view, and therefore will make decisions that could be wrong from a system standpoint. This aspect will be discussed in a few later examples.

Since the OAM&P entity itself is a managed entity and needs to be addressable, it has its own IP address that may or may not be related to the host router's port or forwarding addresses. As such, it can be accessed through internal or external physical or logical lines or channels ("inband" and "out-of-band"). Mostly, the OAM&P cards will have their own separate Fast Ethernet port, connecting to it the NMC or local control console. In some cases, the OAM&P cards communicate using inband signaling, and are connected through the switch fabric. This is of course to some extent undesirable because OAM&P communication in that case relies on functioning links between the OAM&P cards and the switch fabric, and on correct operation of the switch fabric. The protocols that OAM&P cards use to communicate with its managed entities are mostly proprietary protocols, but the protocols and data structures used to communicate with the NMC are standardized. Mostly, SNMPv3 or Q_3 is used for managing network elements on an abstract object model. Software to perform NMC functions is readily available, and sample MIBs help set up nodes and entire networks.

If the system (router, switch, and any other node) has a local maintenance panel with a display and switches, then the OAM&P entity will manage it. The local control panel is exclusively read and driven by the OAM&P entity. Especially in large installations, a local control panel helps avoid confusion so that the service technician works on the intended router and module.

It is not immediately apparent why advanced router architectures should incorporate sophisticated OAM&P functions. After all, current IPv4-based router networks have performed in an acceptable manner so far. While that is true, new requirements have emerged. A best-effort approach is sufficient for unmanaged entities and networks with a typical IPv4 router, but it is not acceptable for a network with IPv6 routers that have to fulfill SLAs. More importantly, it is not good enough for routers that are located in the core of the network. Future advanced routers will have to self-diagnose many more potential problems than they do today, and they must deploy redundant units to ensure continued operation even during failure of one of the crucial components. As such, they are under the supervision of an NMC. The NMC is monitoring all aspects of the routers' uptime and availability, down to the individual module. It does this through the router-internal OAM&P entity. OAM&P, together with appropriate software in the NMC, will enable network administrators to efficiently route traffic. Traffic engineering is only possible with OAM&P functions that support collection of traffic data to determine network load and bottlenecks, and eventually circumvent them (see Figure 10.1). Since neither the originating traffic (the "offered" load) nor the capacity are constant, traffic engineering is highly dynamic and requires fast and efficient statistic traffic data collection as well as appropriate responses to changes in traffic patterns. More importantly, since module failures are unavoidable, some entity must be able to maintain an overview of the operational status of each module and make decisions based on the module status to reroute and redirect traffic in case of a module outage. Even better would be a system that can prevent failures by monitoring error rates

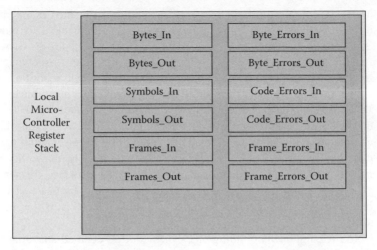

FIGURE 10.1 Register stack in local control CPU.

and making recommendations as to when a module or subsystem should be replaced. Traffic Management and Engineering as well as SLA compliance and High Availability are possible only with sophisticated OAM&P entities and hardware support.

OPERATIONAL STATUSES

Each component, building block, or module—be it a logical or physical entity—can be in exactly one operational status at any given point in time. It must be in exactly one of these statuses. Some of the statuses are reserved for redundant components, building blocks, or modules; others are valid for redundant and non-redundant components, building blocks, or modules. These operational statuses are used to make sure that the device stays operational as long as possible, but they also allow for maintenance of components. The operational statuses of each entity are reflected in a local database within the OAM&P entity, and by a command sent out by the OAM&P entity to the logical or physical component, building block, or module subject to the operational status.

- **ACT:** A component, building block, or module is deemed active by the OAM&P entity if the component, building block, or module is allowed to send and receive data. It is assumed to be correctly functioning.
- **STB:** A component, building block, or module is deemed standby by the OAM&P entity if the component, building block, or module is allowed to send and receive data; however, the receiving side typically is set up such that it ignores the data sent by the entity placed in STB. It is assumed to be functioning correctly. Standby is an operational status that can only be assigned to a component, building block, or module that is one of a redundant pair or triplet. A non-redundant component cannot be in the administrative status standby.

- **MBL:** A component, building block or module is deemed maintenance blocked by the OAM&P entity if the component, building block, or module is not allowed to send and receive data. It is not assumed to be correctly functioning. Its hardware may be functioning according to specification, but its software is either in the process of shutting down, rebooting, or it is being configured with the required configuration data and table entries.
- **DEF:** A component, building block, or module is deemed defective by the OAM&P entity if the component, building block, or module is not assumed to be functioning correctly. It is not allowed to send and receive data; however, the OAM&P entity must assume the defective entity does not obey the OAM&P commands, and therefore may be still sending invalid data. As a result, it will instruct all possible entities that might receive data from the defective entity to disable the receivers that pertain to it.
- **UNA:** A component, building block or module is deemed unavailable by the OAM&P entity if the component, building block, or module is assumed to be not present. It is not allowed to send and receive data; however, the OAM&P entity must assume the unavailable entity does not obey the OAM&P commands. The OAM&P entity will instruct all possible entities that might receive data from the unavailable entity to disable the receivers that pertain to it.

STATUS TRANSITIONS

It is important to understand that status transitions must not be arbitrary. The OAM&P entity must allow certain transitions, but some must be denied. For example, an error can occur at random times to any entity in any status except UNA. As a result, any entity can transition from ACT, STB, or MBL to DEF. An entity in status UNA can only transition to MBL. In hot-standby systems, ACT and STB entities swap operational statuses. In that case, the command to transition to STB must be sent only to the ACT entity, and the STB entity will automatically be transitioned to ACT. If there is no hot-standby entity, the command will be ignored and rejected and an error sent to the NMC.

Allowed status transitions are shown in Figure 10.2.

EXAMPLE 1 FOR STATUS TRANSITIONS

A non-redundant managed entity is in status ACT when a series of unrecoverable errors occur, and the same error type reoccurs persistently. As a result, the OAM&P card takes this particular device out of operation by tagging it as DEF. At the same time, a notification is sent to the NMC requesting service to the device. The service technician arrives and removes the device. It therefore is deemed UNA. A replacement part is installed, and the device status changes to MBL. While it is in status MBL, this device is booting up and verifying normal internal operation. Once it has finished doing so, the service technician can provide the device, if it is not configured automatically, with all required configuration data and databases for its normal

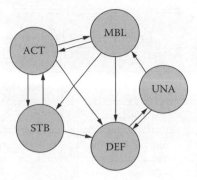

FIGURE 10.2 Allowed status transitions.

operation. As soon as this is finished, the OAM&P card can restore normal operation by changing the device status to ACT. The system is back up again. As is apparent, the MBL phase is important and required to provide configuration data to the replacement module. A transition from DEF or UNA to ACT is therefore not useful and not allowed.

EXAMPLE 2 FOR STATUS TRANSITIONS

A redundant pair of modules in a system is working in a hot-standby configuration, with one module ACT and the other in STB. A timed switchover occurs every 24 hours, triggered by the OAM&P card. The module in STB receives a status change command to ACT, and at the same time the formerly ACT module receives the corresponding status change command to change its operational status to STB. The same commands of the status change are sent to all cards that receive data from the ACT and the STB modules. As a result, all cards that receive data from the redundant pair of modules ignore data received from the STB module (previously the ACT module), and forward the data received from the module that is now ACT (and was STB before the status change). No change occurs in the redundant pair of modules except that the operational status entry was updated to reflect the new status.

RELATIONSHIP WITH NMC

The NMC has the network-wide view of all nodes. It has the current status of all nodes in the network, including failures, outages, and network and node overload and error rates (see Figure 10.3). As a result, the NMC is always in the position to override local OAM&P statuses in case of conflicting information. This is especially important if the network has line outages and the nodes cannot resolve the status of the lines. If the lines are out, then the ports on the routers and nodes cannot determine if the port card has a port-down situation or if the line is down. Only the NMC has messages from both nodes indicating problems with port cards connected to the same line, and therefore the NMC can derive that the line is the problem and instruct the nodes to not shut down the port cards and change their status to DEF or UNA.

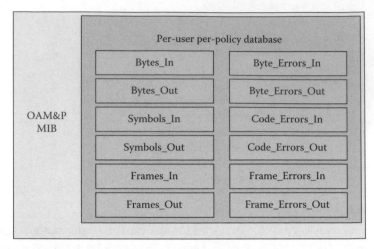

FIGURE 10.3 OAM&P MIB entries.

As a result, the nodes will keep the port cards in ACT; therefore the port cards will be back in fully functional status once the lines are back up.

The OAM&P card sends its nodes' status to the NMC. It sends all status changes to the NMC as well. It changes the operational status of its nodes' subsystems autonomously unless directed otherwise by the NMC.

IMPLEMENTATION

As we have seen, all OAM&P functions are crucial in a router or a switch. These functions are implemented mostly in a centralized OAM&P card, in conjunction with the register sets in all modules. These register sets are located in MACs, HSSL transceivers, framers, network processors, traffic managers, and inside the switch fabric chipset. To reduce the number of registers, a small CPU or microcontroller is usually added to the chip to be able to preprocess the collected data. An ARM or 32-bit MIPS core is sufficient, and even an 8051 derivative is sufficient. This building block can remain hidden from users. A microcontroller or a lower-performance CPU, together with some DRAM and Flash ROM to maintain the required flexibility in interpreting the standards, provides for a very flexible solution that supports OAM&P at a very low cost. These building blocks do not take a lot of space on-chip or on the board, but they provide a significant benefit to the user.

These registers are used to count bits, octets (bytes), datagrams, cells, frames and headers as well as all detected lost bits, corrupted cells, frames, packets, headers, and any other errors that might occur. They allow for an OAM&P entity to easily determine the device-global module-specific Bit Error Rate, system availability, datagram and cell or packet loss rate, and Cell Loss Priority (CLP). Consequently, the OAM&P card can verify local compliance with instituted SLAs. Additionally, the OAM&P card will be able to communicate with the Network Management Center (NMC) and provide device-global data so that the NMC can determine global compliance with the SLAs. However, data collection for OAM&P occurs locally. All individual modules in a communication device, terminal device, or node will collect this statistic traffic data. The OAM&P card's processor periodically polls their results. The data is then compiled and a few crucial Traffic Engineering parameters—as mentioned above—are derived. These are sent to an NMC so that this centralized entity can determine which paths are overloaded, underutilized, too expensive, or out of service because of defects or for maintenance. The NMC then derives appropriate counter-measures from the data it received. Because it has a network-wide overview, it can reroute traffic according to Service Provider specifications and requirements. The OAM&P processor locally installed in each router or switch cannot do this because it does not have the overview. The requirements for traffic rerouting can be SLA and QoS level fulfillment, price or cost of the transmittals, the number of hops, the number of nodes involved, the physical or logical length of paths, BER considerations, or any combination of the above. This is crucial to keep cost down and availability up, as well as to achieve SLA requirements.

FAIL-SAFE AND FAULT-TOLERANT OAM&P ENTITY OPERATION

One might argue that the OAM&P card for the most part does nothing, and therefore its functions can easily be implemented in software as a single task on one of the devices that has spare compute capacity.

This is not a correct assumption. First, the OAM&P card collects status data continuously, in preset intervals. Second, it connects the node (router, switch, firewall within a router) to the NMC. Third, it does not matter when other components fail but, especially during the outage of a card, the OAM&P card is absolutely crucial to the system staying online. It must not fail due to an unknown error condition. Although the OAM&P card is crucially required only during less than 1% of the lifetime of a node, it must not fail during these periods. Therefore, its software must be bulletproof—and it must be independent of the managed entities. It will collect vital statistic data, and these data sets may or may not be corrupted. They may or may not make sense. They may or may not be contradictory or mutually exclusive. It therefore is dangerous to assume that the OAM&P card can operate on valid data. If the communication between the cards and the OAM&P card is inband—which is not recommended anyway—it might not even be able to collect any data at all. For that reason it must be ensured that the OAM&P card can work with data that is entirely unavailable, partially unavailable, missing, or invalid, without crashing. It has to be able to come to a conclusion based on the data it has, independent of whether it is able to verify its validity. The more deteriorated the system status, the more urgent the message to the NMC.

OAM&P ENTITY INTERNAL COMMUNICATION

It is important for the OAM&P entity to be able to communicate with all internal subsystems of the node (switch or router). It must be able to send messages and receive status data back from the line cards, the switch fabric cards, and all sensors and switches. It should also be able to exchange status information with a second redundant OAM&P entity if it is present. The OAM&P card can use data path functions or control path functions within the data path. However, if the data path is down or the control path within the data path is down, then OAM&P becomes impossible. As a result, OAM&P communication within the router should be carried out through dedicated communications channels. It has been proven very cost-effective and more than sufficiently reliant to use Fast Ethernet as a means of communication between the OAM&P entity and the line cards, the switch fabric cards, and in between potentially redundant OAM&P entities. Communication with fan trays and temperature sensors within the chassis is typically implemented with I²C. It must be said one more time that the backplane or the midplane is passive. Therefore, the OAM&P card will not be able to directly query the status of the backplane or midplane.

SOFTWARE IMPLICATIONS

IPv6 and all other advanced protocols that imply Service Level Agreements with QoS parameters have a profound impact on the router software. While at first it might appear that this only has an impact on the line card software or, more generally on modules that are in the data path, the impact of software is not limited to the software on the line cards or the port and processor cards. It has an impact on the software of all entities that are in the data path and outside of the data path. For example, the requirements for the availability of the system directly affect the OAM&P hardware and software. If there is no High Availability requirement, the OAM&P function can be implemented in software on the processor card or anywhere else. However, if there is a need for centralized administration and for HA, then not only does the router require an OAM&P card, it must be able to perform functions that can reset, reboot, and restart any of the line cards, port cards, processor cards, switch fabric cards, and itself. It also must be able to deal with redundant OAM&P cards that can mutually monitor each other's behavior, and take action if something goes wrong. In other words, the software even on the OAM&P card must be more robust while simultaneously must perform functions that the OAM&P functions in other devices do not even have to perform.

First of all, the OAM&P software must support the hardware within all modules and components in its functions. More importantly, it must do so without sacrificing throughput, delay variation, latency, and robustness. It is of no use if the hardware by itself is crash-proof, but the software running on it frequently crashes or must reboot due to software faults. This means that the software must be capable of surviving some hardware errors, but more importantly, it must have a flawless stack and heap administration. Garbage collection must be flawless. In case something goes wrong anyway—like spurious hardware faults or nested ISRs that terminate because of run-time issues—a planned shutdown and restart (recovery) must be implemented. However, rolling recoveries must be avoided.

The impact of software is much more dramatic than most architects, designers and engineers assume. The software boots through a pretty basic mechanism that requires only a small subset of the hardware to properly function. Once it has begun to boot, it will subsequently make use of the present functional detected hardware.

By doing this, the software can start the system by bootstrapping. The software is responsible for providing the hardware with configuration data such that the hardware can perform the desired task—with or without user input. It also deals with exceptions to the anticipated behavior not handled by the hardware, including unanticipated hardware behavior (hardware faults, DRAM ECC errors, I/O errors, link layer errors, CRC errors, and other types). It must deal with errors of the software, including page faults, timeouts of the watchdog timer, stack and heap overflow, failed garbage collection routines, subsequently invalid stack pointers, unanticipated branches and conditions, data format errors, and other mishaps the system might encounter. The hardware may be as good as it gets, but if an error condition in the hardware is not handled properly in software, then the system suffers from this hardware outage, which might cause even more severe crashes of the software. Therefore, it is absolutely crucial that hardware is designed such that errors are reported; all reported errors could be polled by the software to find out what went wrong and what can be done to remedy the situation. Hardware must provide for registers in which the cause of the error and its severity—and possibly even the impact of the error—are documented and recorded such that an Interrupt Service Routine or other method used by the software can be invoked to retrieve evidence of the error and its impact upon the operation of the entire system. If properly done, the software in the subsystem can retrieve the evidence of the error, notify higher layers of software or the OAM&P card, and wait for commands to remedy the situation.

Garbage In, Garbage Out (GIGO) has been and continues to be a fundamental problem in computing. Traditionally, every engineer tries to reduce the "Garbage In" part as much as possible. However, this is unfortunately not possible for the OAM&P entity. By definition, it is most crucial to extract useful information out of the data from a card that is presumed to be defective. For as long as a managed entity works flawlessly, the OAM&P entity only collects the accumulated statistic traffic data and maybe billing information or other policy information. Once the managed entity is not functioning correctly any more, the OAM&P entity must find out how much data traffic—user data or signaling data—is affected, and if this status warrants switching over to a redundant unit or, worse, taking a non-redundant unit out of operation. In other words, it is expected that the OAM&P entity receives some degree of corrupted data from the managed entity, and it must process this data without generating even higher levels of "garbage" data. The matter-of-fact assumption of the GIGO principle being not only valid, but predominant, makes it a lot more complicated on the OAM&P card. The fundamental issue of the OAM&P card is that it by definition cannot know if it can rely on the data in case of a malfunction. The nature of the malfunction is that things do not work right, and therefore it cannot be known if the data received by the OAM&P card is valid or not. While in traditional systems the above-mentioned GIGO problem can be resolved by preventing garbage (or potentially invalid data) coming in, such is not the case for the OAM&P card. The opposite is true, and the system designer must accept this fact. As a result, he must write software on the OAM&P card that can deal with data that is possibly corrupted, invalid, or not available at all. The secret is to extract everything possible, make some assumptions about the validity of the data, assign probabilities to its correctness, and then draw conclusions.

All current and future routers contain a significant portion of software that to a certain degree ensures the flexibility of the routers. Whereas in typical Class 4 and 5 Central Office switches, the design of software has been turned into the science of engineering with a reproducible quality and performance, the router industry is not yet that far. The enormous growth of software in a typical router is mostly to support the ever-increasing number of protocols, increasing inband communication between routers for security and privacy reasons, to perform traffic engineering, and of course to support the growing number of different port types. As a result, an advanced router today contains more software than ever, with a test case coverage that is lower than ever. This will result in a higher number of bugs encountered by the user in the operating system. The rate at which bugs are present in the code will stay the same: one minor bug per 100 lines of code, and a severe bug per 10,000 lines of code. However, if more portions of the code are actually executed, the occurrence of faults will increase. This becomes especially true in inter-processor communication (IPC), when messages can be misinterpreted, access to shared devices using semaphores is not secured, and when task time is exceeded and a hardware task switch is triggered during an incomplete cycle that should otherwise be uninterruptible.

As a consequence, the operating system and the application software, especially on the OAM&P cards but also on all line cards and on the switch fabric card(s), must be written such that it can restart itself through the use of a watchdog timer. The watchdog timer typically is set such that it will trigger a reset on the CPU if it is not reset within its expiration period. Typically, the timer will be set to expire after 512 ms, which typically equals 512 "timer ticks." The software should reset the timer every 500 ms, which it will do if it is not stuck in an infinite loop. If it is, then the main process that resets the timer will not be able to do it, and the software will be restarted by an external reset event through the timer. This carries the potential problem of a rolling recovery—the CPU reboots, and gets into the same infinite loop again, cannot reset the timer, and is reset once again. Events like that are called "rolling recoveries" and are highly undesirable. Unfortunately, they can occur, since it is impossible to write software that avoids them under all circumstances. It can happen that external parameters and conditions are unforeseen or may be out of range, and therefore the same erroneous environment will re-emerge every single time. However, this is a very low probability.

These rolling recoveries on any component in the data path, but also on the OAM&P cards, must be avoided at all cost. Therefore, not only must the system test be very thorough, but more importantly, to allow for testing a significant portion of all system statuses and status transitions, the system must be set up such that each and every individual module is as simple as possible. This, for the most part, forces a distributed design with distributed hardware and software instead of a monolithic hardware and software design. To achieve this, one possibility is to design the system with the idea in mind that dedicated functions are performed by dedicated software on dedicated hardware. This approach allows for displacement of large and complex hardware components using large and complex software by small modules running comparably simple software. This is true for the OAM&P cards and switch fabric cards, but most importantly for the line cards that must process data path datagrams—for protocols, queuing, discarding, switching, and for inband communication. In fact, the line cards typically will contain the most complex hardware and software. This is to a certain degree true even in midplane architectures, where the PHY, the MAC, and the framer are on the port cards and the processing occurs on the processor cards. The OAM&P entity cannot resolve rolling recoveries on a line card if the fault condition re-emerges every single time the line card has finished starting. The local control processor will receive the reset signal from the OAM&P entity and restart the line card. However, if the conditions remain the same, then the line card will not ever get past MBL status, and it will repeatedly fail at the same point of the software. As a result, the only action the OAM&P entity can take is to place the line card into DEF and notify the NMC. The OAM&P entity cannot resolve persistent software faults on the line cards or the switch fabric cards, and therefore does not replace and should not reduce a thorough system test.

As has become apparent, software exception handling on the OAM&P card is crucial. The OAM&P entity must respond to exceptions as soon as possible, and must be able to poll the critical status from the originating entity. This can be quite a challenge. As explained earlier, the arrival rate of OAM&P packets can be quite high. Most of them will be for regular status updates, and perhaps for billing information, but once error messages come in, they must be processed immediately. The priority of these messages is higher than those of regular messages. Not only must the OAM&P entity take into account the different priorities of these messages, it must process them accordingly.

Software is affected dramatically on all levels by the requirements for a modern router. This is the case for software in the data path, but more importantly for the OAM&P software. While the software in the switch fabric card probably does not perform any data path functions, it must be aware of the health and administrative status of the switch fabric because it needs to notify the OAM&P card of the necessity of a switchover and execute it if it receives an appropriate command. Whether the switchover is lossless or not is determined by hardware, by OAM&P software, and by the urgency and the specific circumstances of the switchover request. The software on all components—whether in the data path or not—in an IPv6 router is more substantially impacted by the increasingly stringent requirements of SLAs.

ENCRYPTION FOR NMC-TO-OAM&P TRAFFIC

An additional burden will be placed on the OAM&P entity once encryption of messages and commands has become mandatory. In this case, the OAM&P entity will have to encrypt and decrypt messages and commands, and it will have to access public keys within the Public Key Infrastructure (PKI). In the future, we will see additional traffic to PKI servers for authentication of messages. The deployment of PKI to communicate with the router will be dependent on the spread of malicious software that tries to attack the core network. The faster an attack spreads which undermines manageability of routers in the core network, the faster PKI will be used as a means to encrypt data exchanged between the NMC and the router OAM&P card. This will require some additional computational power in the OAM&P card, but it is manageable by today's CPUs and microcontrollers.

More importantly, the keys must be stored in a way that an intruder cannot access them.

EXAMPLES OF FAILURE MODES OF MANAGED ENTITIES

A router with redundant switch fabric cards, two OAM&P cards, and line cards connected through HSSL using 8B/10B encoding will serve as an example. Assume each switch fabric card and each line card has a local processor to boot up the cards and to monitor card health. One of the switch fabric cards starts encountering increasing levels of 8B/10B code errors. It gathers evidence in the counters for the code errors and sends a message to the OAM&P cards stating that within the last 15-minute interval, there were 200 code violations. In the previous interval, there were 50. The OAM&P cards receive the message and compare that to historical values. They poll the fan statuses and query the line cards for code errors. They also compare the code violations of switch fabric card A to switch fabric card B. Without knowing the comparative numbers, a decision cannot be made. The other switch fabric card had sent messages prior to that point in time and indicated that its level of code violations is even higher and had been higher for awhile. Additionally, all line cards report code violations after having been polled, and the fan tray status is that tray 1 shows all six fans non-functional, and tray 2 has only two out of six fans working. Therefore the assumption must be that something much more serious is going wrong, and the switchover would only make things worse. The only real solution is to signal to the NMC that service is required on this router as soon as possible. A truck roll will have to be initiated. Software on the switch fabric card, the line card, or the OAM&P card that mishandles the situation would have brought down the router earlier. If the local control processor were allowed to reboot the system without a command from the OAM&P card, the scenario would have been worse. In that case, the switch fabric card would have been restarted by the OAM&P card, causing a lossy switchover from the active to the hot-standby switch fabric card. However, since it would have been a lossy switchover, the line cards would have encountered even higher rates of code violations, and would have started to reboot themselves. This would have caused more code violations, essentially bringing down the switch fabric cards. Both the active and the hot standby card would have encountered enormous numbers of code violations, and the only logical conclusion to draw out of this local data would have been to reboot and switchover. If the decision to reboot were made truly locally, even in the switch fabric cards, both of them would encounter code violations, conclude that a switchover is required, and initiate a switchover and reboot. Not only would that bring down the switch fabric cards—because both would be in rolling recovery—it would also bring down the line cards since they will not be able to transmit anything through any of the two redundant parts of the switch fabric. Any data they send is discarded, and both ingress ports from the active and the hot-standby switch fabric card receive invalid data and so will try to reestablish the links, with no avail. The statistics counters for discarded datagrams will go up since now almost 100% of all transmittals are invalid and therefore encounter code violations on both ingress ports. This is true for all line cards and both halves of the switch fabric. At this point in time, it is impossible for an OAM&P card to figure out where the problem began, and even if the OAM&P card has reasonable software to remedy problems, it cannot do much. All gathered evidence at this point in time is worthless, nor does the fan tray status provide any additional or good information.

It therefore is essential that all cards have local intelligence to gather information and evidence for root-cause analysis of the problem, but that they cannot and must not initiate any actions without the OAM&P card sending commands to perform a local action. This is true even for a link-down situation because the link involves two sides. Only on truly local events may the local processor take action. It may read a DIP switch and display the result on an LED. It may also display the current device status that was assigned by the OAM&P card on a local LED. All other actions other than bootstrapping must be executed only upon messages from an OAM&P card. As a consequence, the OAM&P card of course must be protected, and it must not catastrophically fail without the redundant (hot-standby) OAM&P card taking over. This is not an easy task.

To clarify this, let us assume we have a router with a switch fabric card and line cards. Furthermore, the communication between line card L1 and the switch fabric card is experiencing unusually high error rates. None of the modules can determine where the errors occur. They can originate in the driver IC for the line card on the switch fabric card, the backplane or midplane, the connectors between backplane or midplane and switch fabric card or line cards, or in the driver IC of the line card towards the switch fabric card. If all of these modules are allowed and able to determine and set the operational status of any of those modules, each module would have taken itself out of operation, changing the module status from ACT to DEF. Since the backplane and midplane are passive and therefore do not have active components, their operational status can only be derived or deducted through exclusion. However, a backplane or midplane in status DEF brings the entire router down, since it is disabled. This would not be a good choice. As a consequence, the OAM&P card must be programmed such that it takes components or modules out of service in the order of least impact first and most severe impact last. It therefore would first assume that the line card might be defective, and bring it into the operational status MBL. This would finish all current transactions and start rejecting new traffic. After awhile, no traffic goes through that line card anymore, and a maintenance person can replace the line card with a new one. Pulling the old presumed defective card out of the system renders its operational status UNA. Inserting the new card brings it into MBL first, so that the OAM&P card can trigger the line card's bootstrapping procedures to start it and supply all relevant metadata, including routing tables and so on. During the bring-up, it will begin to communicate with the switch fabric card and will start to count transmission errors. If none occur, then indeed the line card was defective, and the problem is solved. The OAM&P card can then change the operational status of the line card to ACT. The last part of the sequence is true if the card is malfunctioning and has found an internal line-card problem. In this case, it would report this to the OAM&P card, and it would be set to DEF. After replacement, the bring-up procedure is the same as mentioned above.

If the problem persists, then the errors were not due to defective components on the line card. The error must therefore be on the backplane or midplane, the switch fabric card, or in any of the connectors. If the router has a redundant set of switch fabric cards, then a switchover should clarify if the switch fabric card is the culprit or not. If the switch fabric cards operate in hot standby, then the OAM&P card must just bring the active card out of ACT and into STB, and at the same time bring the card that was in hot-standby into active ACT operational status. This simple switchover should not result in data loss at all. If the problem is subsequently solved, then the now-STB switch fabric card was the culprit. The OAM&P card then changes its operational status into DEF to alert the NMC that it needs to be serviced. The service personnel must swap the defective switch fabric card for a functioning one. Removing the defective card brings its operational status to UNA, and inserting the new card brings its operational status to MBL. It can then be brought up and provided with all necessary configuration data. Once it is fully functional, it can be switched to STB.

However, if that does not solve the problem, then the source of the error lies in the backplane itself, or in the connectors. This is an event the user always tries to avoid, because the backplane (and the same is true for the midplane) is bolted into the chassis. A defect in the backplane or the midplane means that all of the cards must be removed from the chassis, all cables disconnected, and the chassis removed from the rack. A new chassis is installed, all cards are reinserted, and their cables reconnected. The entire system has to be brought up and provided with all required setup and configuration data.

This is a time-consuming and therefore very expensive proposition, because time and money are spent and the device is out of operation and therefore does not generate revenue. Even if traffic can be rerouted and the SLAs are not impacted, this is a highly undesirable situation.

THE NECESSITY FOR THE DEVICE-GLOBAL VIEW

One reason the OAM&P function must be device-global and not local is the fact that only an entity that has a device-wide view can make decisions that weigh conditions properly. Let us assume that in the above example one switch fabric card is ACT and encounters constant errors on all line cards, with an error rate exceeding normal operations but not bad enough to warrant a line-down situation on any of the High Speed Serial Links. If the switch fabric card were in charge of determining its own operational status, it would switch over to the hot-standby switch fabric card, and declare itself DEF. The hot-standby switch fabric card receives the same datagrams and performs the same switching operations, therefore outputting the same datagrams to the same line cards as the ACT switch fabric. However, the line card that is the designated uplink card for this router receives 100% bad datagrams from the hot-standby switch fabric card. As a result, this switch fabric card will put itself out of operation as well, by declaring itself DEF. As a result of local OAM&P on each of the cards, either both switch fabric cards would be in DEF, or it would depend on the timing as to which one would be ACT and which one would be DEF. Such a probabilistic behavior is not acceptable, especially since one choice would effectively shut down the entire router—a 50% probability the router would be shut down completely. In the case of a device-global OAM&P card, the card would have the information available about each of the data path cards (and the hot-standby redundant OAM&P card, if in a redundant system), and would make the obvious right choice; it will leave the switch fabric card with the raised error rate in ACT, put the hot-standby switch fabric card in DEF, and alert the NMC. Suspected defects in the hot-standby switch fabric card and the uplink card will be signaled too. A service technician will be dispatched to change the switch fabric card, and then the error rate on the datagrams sent to the uplink card will be monitored. If the line errors and code violations disappear, the switch fabric card was defective. If the problem persists, then the switch fabric card was not the root cause of the problems, and the cycle has to begin anew with a new suspected defective part.

For example, a line card could report it has received 1,000 packets, 500 of which were corrupted. The switch fabric card may have reported that it received 250 packets from this line card, all of them intact. The local OAM&P card therefore could have derived that this particular line card is defective, and that it should send an administrative alert to the NMC. The NMC then would not only notify adjacent routers of the outage, but also send back a notice to take this route out of service, together with the command to reroute locally as much as possible, and send the rest of the traffic using alternative routes with different hops.

SAMPLE OAM&P CARD SCHEMATIC

An OAM&P card in effect is a card with a main processor and means of communications to the router's internal entities in the data path, as well as to all other

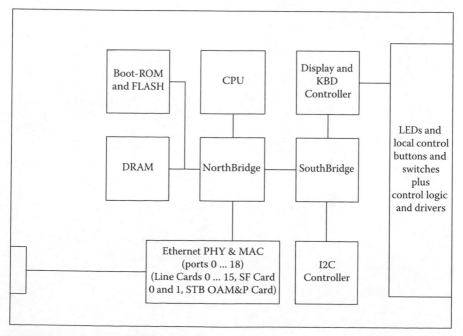

FIGURE 10.4 Sample OAM&P card.

contributors to failure modes. As a result, it will have to contain the logic to connect it to all active cards, via Ethernet or Fast Ethernet and I²C, to all sensors in the chassis or cabinet that monitor parameters requiring supervision. These are the temperature sensors in the chassis and the fans trays, as well as intrusion detection systems at the doors and hinges. See Figure 10.4 for an example of an OAM&P card schematic.

CONCLUSION

OAM&P is a crucial component of advanced router architectures. It is based on registers in all modules and components, as well as on a device-central OAM&P entity that collects traffic data and assigns operational statuses to all modules and components. It is responsible for maintaining operation of the device and for initiating service calls to the NMC. OAM&P requires the managed entities to communicate with the OAM&P entity, preferably out-of-band. As a result, it is extremely important that all OAM&P functions are supported by the hardware—both in the data collection as well as in the local, but device-global preprocessing, of OAM&P data. Data collection for OAM&P often is implemented by providing register sets that support the basic nine groups for the Simple Network Management Protocol version 3 (SNMPv3). It will be almost always a subset of the full Management or Managed object Information Base (MIB) described in that standard. Consequently, High Availability can be achieved. Centralized OAM&P can lead to significant operational cost reductions while maintaining network integrity and security.

Glossary

AAL: ATM Adaptation Layer.

AAL5: ATM Adaptation Layer 5. AAL translates large Service Data Units (such as those produced by IP) into 48-byte ATM cells and vice versa. AAL5 is by far the most important AAL implementation, considered by many as native ATM. In a system, the entity that performs AAL5 function is often called Segmentation And Reassembly (SAR).

ABR: Available Bit Rate. ATM traffic type.

Address Mask: A bit mask used to select bits from an IP address for subnet addressing. The mask is 32 bits long in IPv4 and 128 bits long in IPv6, and selects the network portion of the IP address and one or more bits of the local portion.

ADM: Add-Drop-Multiplexer.

ARP: Address Resolution Protocol. The TCP/IP protocol used to dynamically bind a high-level IP address to a low-level physical hardware address.

AS: Autonomous System.

ASIC: Application-Specific Integrated Circuit. An integrated circuit designed and manufactured for a specific purpose.

ASSP: Application-Specific Switching Processor. An ASSP is an ASIC that performs a certain function or service, based on a processor architecture that is optimized towards the service task only.

ATCA: Advanced Telecommunications Computer Architecture. Based on PCI Express, but with a more compact module outline and a connector that is not a card edge connector, this hot-pluggable derivative of PCI Express has started to make inroads into telecommunication and datacommunication systems. ATCA systems have the ability to work in conjunction with a switch fabric instead of a shared bus.

ATM: Asynchronous Transport Mode. ATM transports, switches, and identifies cells based on their header information at a fixed location within the cell, and not based on a timeslot assignment. ATM is capable of consolidating all traffic types. Gateways are required for services with their own addressing schemes that are not based on E.164.

ATMF: ATM Forum. Defines, accepts, and verifies new ATM standards independently of ITU-T.

BC: Bearers' Capabilities. BC describe the functional capabilities of the terminal device in ISDN.

BER: Bit Error Rate.

BGP: Border Gateway Protocol.

BGP-4: Border Gateway Protocol Version 4.

B-ISDN: Broadband Integrated Services Digital Network.

Bit: Binary digit. Atomic information unit in modern digital information.

Broadcast: A packet delivery system that delivers a copy of a given packet to all hosts that attach to the network.

Byte: Numeral consisting of 8 bits, with the MSB being the leftmost bit in little endian architectures.

CAGR: Compound Annual Growth Rate.

CAM: Content-Addressable Memory. This is an important lookup engine in n-tuple lookups within ATM switches and routers.

CBR: Constant Bit Rate. ATM traffic type.

CCITT: Comité Consultatif International de Téléphone ét Télégraphique.

CCP: Connection and Control Part.

CCS#7:   Common Channel Signaling System number 7.

CDV: Cell Delay Variation. This is an important traffic parameter in ATM. CDV sets boundaries for the variation of the arrival time of cells.

Cell: A datagram of a fixed length, with the fixed-length header in a fixed position within the datagram.

CHAP: Challenge Handshake Authentication Protocol.

CIDR: Classless Inter-Domain Routing.

CISC: Complex Instruction Set Computer. This is a processor core with large number of instructions, which requires a nanocode ROM for the atomic instructions, and needs multiple cycles per instruction execution. CISC processors provide moderate pipelining capability, and they show nearly no ability for (atomic) parallelism. Most network processors are CISC-type CPUs.

CIX: Commercial Internet Exchange and ISPs. The CIX is a point at which the NAPs interconnect ("peer"). Typically, there is one national CIX, and they are all consolidated in the international CIX. Since not all traffic is hierarchical, the international CIX is not more heavily loaded than some national CIXes. Some countries share a single CIX. With growing traffic everywhere in the world, there will be at least one national CIX per country.

CLEC: Competitive Local Exchange Carrier. The CLEC does not necessarily have to rely on POTS or ISDN wiring to the end user (the "last-mile-problem"). They can use Cable Modem access technology, xDSL technologies, or power line communications. Even Wireless Access seems to be a viable approach.

CLI: Common Line Interface.

Client: A client is a device that relies on having services (file services, application services, print services, access right administration services, etc.) carried out by a system that supplies it with these. Some fundamental functions of the client's programs are carried out on the server, so the client typically does not work without an attached server.

CLP: Cell Loss Priority. CLP is a mechanism for advanced routers and ATM switches to determine drop policies based on predefined parameters.

CMIP: Common Management Information Protocol.

CMIS: Common Management Information Service.

CO: Central Office. A Central Office typically contains Class 4 or 5 Central Office POTS switches and some other equipment colocated with it that can make use of the proximity. DSLAMs, ADMs, and routers typically are colocated with CO switches. Usually, all CO equipment is carrier-class equipment, owned by a carrier. CO equipment has much higher demands in reliability and performance compared to office or SoHo equipment.

Compute-bound: A system is called compute-bound when its throughput is limited by the availability of computational performance. The opposite situation is called I/O-bound.

Compute-intensive: Requiring a large amount of computational performance.

CoS: Class of Service. This is a service within IP to distinguish in between different priorities of datagrams.

cPCI: Compact PCI. Based on PCI, but with a more compact module outline and a connector that is not a card edge connector, this derivative of PCI has made inroads into telecommunication and datacommunication (datacom) systems. The broad availability of PCI-based communication IC's has made these systems comparably cheap while maintaining better reliability and mechanical ruggedness.

CPE: Customer Premises Equipment. CPE may be owned by the user or a service provider. Especially ISPs tend to buy the equipment and then lease it to the users.

CPU: Central Processing Unit.

CSIX: Common Switch Interface. Network Processor Forum standard.

CSMA/CD: Collision Sense Multiple Access with Collision Detect. A characteristic of network hardware that operates by allowing multiple stations to contend for access to a transmission medium by listening to see if the medium is idle; also a mechanism that allows the hardware to detect when two stations simultaneously attempt transmission. This describes a method for media access, which is used in Ethernet installations. With this technique, a station which is willing to send, first listens on the common media whether another station is sending, or if there is currently no traffic. If the latter is the case, then the station begins sending the datagram. It monitors the line continuously and compares the sent signal to the received signal. If they differ, a collision has occurred. In this case, sending is interrupted. The resend attempt will then be carried out later, after a stochastically determined period of waiting time. Since the packet of the other station had been destroyed as well, both stations need to resend. The more users or the smaller the packets are, the higher is the likelihood of destructive collisions. The effective data rate first dramatically decreases, and then converges to zero with an increasing amount of collisions. CSMA/CD is necessary because of the shared media structure of Ethernet. Many devices share a single resource. This technique is optimized for a small number of users in a single collision domain. Higher numbers of users must be consolidated using bridges, routers, or switches. The Ethernet topology is a bus. This is not true anymore for Fast Ethernet switches with dedicated ports per user. Here the Ethernet-shared media (the bus) is collapsed into the shared memory, the crossbar, or the switching network of the router or switch. Collisions cannot occur anymore in such a case.

CTI: Computer Telephony Integration. The goal of CTI is to integrate both the PC and the telephone so that the PC can not only carry out a phone call, but the user also could use the contact database in the PC. The first step will be to define an interface between both devices so that the PC can control the phone and the phone numbers can be downloaded from or dialed by the PC.

Cycle Time: Inverse of the clock frequency.

DA: Destination Address.

DDNS: Dynamic Domain Name Service/Server.

DECT: Digital Enhanced Cordless Telecommunications.

DES: Digital Encryption Standard. It uses a 56-bit key to encrypt data. DES is not a very strong encryption, but it can be used iteratively. A more useful application of DES is Triple-DES with a resulting 168-bit key.

DFT: Design For Test. DFT is an important design criterion in large systems, to ensure the system is testable and that the tests cover a significant portion of real-world situations.

DFT: Discrete Fourier Transform.

DHCP: Dynamic Host Configuration Protocol.

DNS: Domain Name Service. A DNS is a name resolution service. It converts a symbolic name into an IP address, e.g., http://www.dekker.com $\rightarrow$ 216.168.228.43.

Downstream: Downstream is the direction of data flow from the server towards the client, or from the node to the terminal.

DRAM: Dynamic Random Access Memory. This is cheap, but comparably slow memory technology that is used in computers, servers, routers, and ATM switches as the main memory. Although it is slower than the SRAM-based cache, it is the predominant main memory technology. Its high density at low prices makes it the ideal solution for large databases (policies, lookup, and billing tables) and other applications in network processors and traffic managers.

DSO: Digital Signal Level 0, or Digital Service Level 0, depending on whether it is derived from ITU-T or ANSI. This is a digital transmission standard for POTS with an 8 kHz sample rate and 8 bit resolution per sample. DSO is based on Pulse Code Modulation (PCM). Often, a timeslot within the PSTN is referred to as DSO.

DSLAM: Digital Subscriber Line Access Multiplexer. A DSLAM connects the Class 5 switches' DSL subscriber ports to an ATM switch or a router.

DSP: Digital Signal Processor. DSPs are used to process signals.

DTE: Data Terminal Equipment.

DTMF: Dual Tone Multi-Frequency. With DTMF, two out of four tones are superposed, giving access to sixteen combinations. This is the Plain Old Telephony System (POTS) end user signaling system. Modems use DTMF.

Dual Ported RAM: Dual Ported Random-Access Memory. This is typically an SRAM with two ports that can be accessed simultaneously by two different entities. It is usually deployed as a semaphore.

DVMRP: Distance Vector Multicast Routing Protocol. Multicast protocol that attempts to replicate datagrams at such point in the network to consume the least amount of resources.

DWDM: Dense Wavelength Division Multiplex. DWDM is used in OADMs to multiplex multiple wavelengths onto a single fiber to increase its channel capacity.

E.164 address: An address type defined by CCITT und ITU-T for POTS and ISDN.

E911: Emergency 911 phone calls.

EBGP: External Border Gateway Protocol.

EGP: Exterior Gateway Protocol. EGP has been replaced by BGP.

EPD: Early Packet Discard. EPD is a mechanism for advanced routers and ATM switches to drop excess traffic as early as possible based on predefined parameters.

FC/AL: Fibre Channel/Arbitrated Loop. FC/AL is a non-routable protocol in network storage systems. FC datagrams can be encapsulated in IPv4 or IPv6, and then routed.

FCC: Federal Communications Commission. The FCC is a TelCo regulatory authority in the US.

FFT: Fast Fourier Transform. FFT converts time domain data into frequency domain data and vice versa. Real-life implementations in DSPs or any other processors are Discrete Fourier Transforms since DSPs operate with discrete fixed-point or floating-point numbers. Fast Fourier Transform results are numerically approximated by results of Discrete Fourier Transforms.

Fibre Channel: Fibre Channel is defined in ANSI X3T9.3, and was later transferred to ANSI X3T11. FC is used mostly to connect FC harddisks to FC adapters in servers.

FIFO: First In, First Out. FIFO is a register stack architecture that maintains the order of incoming datagrams.

FPGA: Field-Programmable Gate Array.

Frame Relay: Network technology that was mostly used in MANs. Frame Relay uses frames to transmit data.

FRF: Frame Relay Forum. Defines, accepts, and verifies new Frame Relay standards.

FTP: File Transfer Protocol. This protocol defines the basic functions for transferring a file from one station to another. This is the originating protocol for the terms, "upload" and "download."

Gbit/s: Gigabit (10^9 or 1 billion bit) per second.

GMII: Gigabit Media Independent Interface. GMII is a standard gigabit PHY-to-MAC interface defined in IEEE 802.3ae.

GPRS: General Packet Radio Service.

GPS: Global Positioning System. GPS is used for location systems. In the US, the FCC has mandated that cell phones can be located for and during Emergency 911 (E911) calls. As a result, newer phones might be equipped with GPS receivers and periodically or upon command reveal their location to the MSC.

GRE: Generic Routing Encapsulation.

GSM: Global System for Mobile Communication.

GSR: Gigabit Switched Router.

GTLD: Generic Top Level Domain.

HA: High Availability. Typically, system availability of more than 99.999% is required for HA designation.

HDLC: High-level Data Link Control.

Header: Part of a datagram that contains information about the sender's address, the intended receivers' address, and some other optional parameters regarding the priority and the importance of the datagram and its specific timing requirements.

Hierarchical Routing: Routing based on a hierarchical addressing scheme. Most TCP/IP routing is based on a two-level hierarchy in which an IP address is divided into a Network portion and a host portion. Routers use only the Network portion until the datagram reaches a router that can deliver it directly. Subnetting introduces additional levels of hierarchical routing.

HMAC: Hashing for Message Authentication.

HomePNA: Home Phoneline Networking Alliance. HomePNA has the potential of linking all home alliances, and connect them or a home gateway controller to the Internet. This is a potentially huge bandwidth requirement for the Internet.

Host: Any computer system that connects to a network and is not a client.

HSBI: High Speed Backplane Initiative. HSBI was an initiative that tried to unify backplane designs so that router manufacturers could chose among vendors for line cards, switch fabric cards, and backplanes for modules. HSBI merged with OIF. HSBI set out to promote HSSLs over backplanes and standardize the format.

HSCSD: High Speed Circuit-Switched Data. This technology allows bundling 9.6 kbit/s cell phone channels and therefore might enable reasonable surfing and searches on cell phones, potentially generating huge amounts of traffic.

HTML: HyperText Markup Language. HTML is the basis for the web pages that make up the majority of the Internet today.

HTTP: HyperText Transfer Protocol. HTTP defines how to transport HTML documents from a server to a client. HTTP is a protocol to transmit world-wide web (WWW) pages.

IANA: Internet Assigned Names Authority. The IANA assigns public Internet addresses which are members of three different classes depending on the volume of requested numbers (Class A, B, and C). ICANN has taken over IANA's responsibility.

IBGP: Internal Border Gateway Protocol.

ICMP: Internet Control Message Protocol.

IEC: International Electrotechnical Commission.

IEEE: Institute of Electrical and Electronic Engineers. The IEEE is an important standards body.

IETF: Internet Engineering Task Force.

IGMP: Internet Group Management Protocol.

IGP: Interior Gateway Protocol. This includes IGRP, OSPF, and RIP.

IGRP: Interior Gateway Routing Protocol.

ILEC: Internet Local Exchange Carrier.

IN: Intelligent Network. IN is a service within the telephony network that relies on CCS#7 and servers for address translation and other services. IN is required for services such as 800 number translation, Local Number Portability (LNP), and other lookup functions that are invisible to the end user.

INAP: Intelligent Network Application Programming Interface (a.k.a. IN-API). It is a library of functions to be used with the Intelligent Network (IN). INAP requires local processing power in the terminal device.

I/O-bound: A system is called I/O-bound when its throughput is limited by the throughput of its I/O channels. The opposite situation is called compute-bound. A system can be compute-bound, I/O-bound, or balanced. The type of boundedness heavily depends on the application. While a system may be I/O-bound in one set of applications, it may be compute-bound in another set of applications, and it may be balanced in a third set. Network Processors used to be I/O-bound, but with the newer HSSLs and the requirements for Deep Packet Investigation (DPI) today, they typically are compute-bound.

I/O-intensive: Requiring a large amount of I/O bandwidth.

IP: Internet Protocol. The IP suite of protocols is a collection of programs that define and enable interconnectivity between two or more stations using either serial lines, WAN links, or Ethernet and other transport mechanisms. The Internet Protocol consists of a suite of protocol for system intercommunications such as SLIP, PPP, and CSMA/CD in the case of Ethernet. IP is a connectionless protocol, in contrast to connection-oriented WAN protocols like PPP, POTS, or ISDN.

IP address: IP addresses are Layer 2 addresses assigned by an administrator, the user, or a dynamic IP server. They can be private (not assigned by the ICANN) or public (assigned by the ICANN). Private IP addresses can be used internally on LANs, but must not be transmitted to the public Internet. Public addresses may be used on the Internet. Private address blocks may be converted to one single public IP address using NAT, if only one public IP address is assigned. In IPv4 both the source address (SA) and the destination address (DA) are 32 bit long, and in IPv6 both the source address and the destination address are 128 bit long.

IPCP: Internet Protocol Control Protocol.

IPng: IP next generation. IPng became IPv6; IPv5 was discarded.

IPv4:   Internet Protocol Version 4. This is mostly a best-effort protocol for data networking.

IPv6:   Internet Protocol Version 6.

IS-IS: Integrated Intermediate System-to-Intermediate System.

ISDN: Integrated Services Digital Network. ISDN was the first attempt to integrate data and voice services seamlessly. ISDN today is the basis for user access (even using POTS) and internode communications. Typically messages between the user and the Class 5 switch or between switches are ISDN Q.931 messages. Messages in between these Class 5 and Class 4 switches in the backbone switches are called CS#7 (a.k.a. SS7) messages; they are transported on separate trunks independent of the data and voice traffic. CCS#7 generates an overlay network. The relevance of ISDN for the user is somewhat limited in the US, while it is very strong in Europe. Since the call setup and the call clearing are very quick, the need for instant-on (always-on) user access to TelCo equipment or Internet backbones does not exist; ISDN helps free resources. The most important issue about ISDN today is not only the integration of the data and voice services; it is much more the possibility to provide medium-speed worldwide data transfer over existing networks and LAN interconnectivity. ISDN basically comes in two major versions. The Basic Rate Interface (BRI) offers, in addition to the signaling channel with 16 kbit/s, a data transfer rate of 64 kbit/s (US: 56 kbit/s) per channel. ISDN allows for the possibility of bundling channels, giving even 128 kbit/s (US: 112 kbit/s) in total for BRI. The Primary Rate Interface (PRI) provides 30 (US: 24) data channels and one (US: zero) signaling channel with 64 kbit/s (US: 56 kbit/s) each. Depending on the capabilities of the router hardware, 2–30 (US: 24) channels can be bundled with PRI. So with PRI the user has a total data transfer rate of up to 1920 kbit/s (US: 1544 kbit/s). The according standards are Q.931 and Q.2931.

ISO: International Standardization Organization.

ISP: Internet Service Provider. An ISP provides the customers with Internet access and services related to the Internet: mostly email hosting, IRC, and Instant Messaging. Some ISPs run spam filters and virus checks on customers' emails. Rarely will an ISP provide SLAs to its customers (yet).

ISUP: ISDN User Part. ISUP describes the communications parameters, signaling, messages, and commands between the ISDN Central Office Switch and the ISDN Terminal Device.

ITSP: Internet Telephony Service Provider. An ITSP is a service provider that offers its clients telephony services based on VoIP and FoIP. These services can be POTS or ISDN to the client and packet services in the backbone, or they could entirely be based on TDM over packet.

ITU-T: International Telecommunications Union—Telephony Section (the successor to CCITT). This is the international standards body for hardware, software, protocols, messages, and commands related to telecommunications.

Kbit/s: Kilobit (10^3 or 1 thousand bit) per second.

LAN: Local Area Network. A LAN typically is limited to a building or a campus. Connecting LANs via a WAN usually requires a router and introduces the necessity for a gateway. LANs operate at 10 Mbit/s (Ethernet), 16 Mbit/s (Token Ring), 100 Mbit/s (Fast Ethernet, FDDI), 155 Mbit/s (ATM STM-1 or OC-3), 1000 Mbit/s (Gigabit Ethernet, GbE), and 10 Gbit/s (10 GbE).

Layer 1: A reference to physical-level communication (e.g., the MAC address) or physical connections derived from the ISO 7-layer reference model.

Layer 2: A reference to link-level communication (e.g., frame formats) or link-level connections derived from the ISO 7-layer reference model. For LANs, layer 2 refers to physical frame format and addressing.

Layer 3: A reference to network-level communication derived from the ISO 7-layer reference model. For TCP/IP Internets, layer 3 refers to IP and the IP datagram format.

Layer 4: A reference to transport-level communication derived from the ISO 7-layer reference model.

LCI: Local Connection Identifier. LCI is used locally to route a datagram. The LCI is determined from headers and sometimes the payload of the datagram. It is not used outside of the device.

LCR: Least Cost Routing. LCR is a routing protocol that uses billing information to route traffic.

LD: Long Distance (LD call, LD carrier, etc.)

LDAP: Lightweight Directory Acccss Protocol.

LDP: Label Distribution Protocol.

LIC: Line Interface Card.

LNP: Local Number Portability. See IN.

LSB: Least Significant Bit. The bit order in datagrams may be little endian or big endian. In one case the LSB is at the right-most position of the byte or datagram, in the other it is in the left-most position of the byte or datagram.

LVDS: Low Voltage Differential Signaling. LVDS is a robust signaling system for high-speed signals between integrated circuits.

LVTTL: Low Voltage Time To Live. This is a TTL-type signaling with smaller signal swing.

MAC: Media Access Controller/Media Access Control. The MAC part is Layer 1 on OSI ISO model. A general reference to the lower level hardware protocols used to access a particular network. The term MAC address is often used as a synonym for physical address.

MAC address: The MAC address is a unique address in a MAC address ROM within any NIC. By using the MAC address a sender or a destination can be identified on the MAC (=physical) Layer of the OSI ISO model.

MAN: Metropolitan Area Network.

Mbit/s: Megabit (10^6 or 1 million bit) per second.

MBONE: Multicast Backbone.

MD5:   Message Digest 5. Encryption standard.

MESI: Modified, Exclusive, Shared, or Invalid. This represents the four statuses in which a cache entry of a Multi Processor System or a Massively Parallel Processor System may be. It does not apply to single-processor systems. Each cached entry might either be Modified, Exclusive, Shared, or Invalid. The MESI protocol is used to ensure cache coherency. Many Class 5 CO switches and some routers deploy ccNUMA cores with MESI within their SMP cores.

MFC: Multi-Frequency Code.

MFC-R2:   Multi Frequency Code based on superposing two out of four frequencies.

MIB: Management Information Base (in SNMP). A router maintains this set of variables (database) if it is running SNMP. Managers can fetch or store into these variables. MIB-II is the current standard.

MIB: Managed object Information Base (in ITU-T).

MII: Media Independent Interface. MII is a standard PHY-to-MAC interface specified in IEEE 802.3u.

MIPS: Mega Instructions Per Second. Since this does not determine how powerful these instructions are, or if they are dependent one on each other or exclude pipelining in certain conditions, the MIPS value does not say too much. MIPS values are a result of the benchmark test suite. SPECint shows results which are nearer to real life.

Modem: Modulator/Demodulator. Typical analog modems over POTS provide a baudrate of 56 kbit/s; DSL modems typically have a downstream data rate of 1.5 Mbit/s and up to 384 kbit/s upstream; G.SHDSL modems have symmetric bitrates of up to 44 Mbit/s.

MPLS: Multi Protocol Label Switching.

MPoA: Multi Protocol over ATM.

MSB   Most Significant Bit. The bit order in datagrams may be little endian or big endian. In one case the MSB is at the left-most position of the byte or datagram, in the other it is in the right-most position of the byte or datagram.

MSC: Mobile Switching Center. The MSC is the equivalent of the Class 5 Central Office switch for cell phone traffic and infrastructure.

MSDP: Multicast Source Discovery Protocol.

MTP: Message Transfer Part.

MTP-L3:   Message Transfer Part, Layer 3.

Multicast: A technique that allows copies of a single packet to be passed to a selected subset of all possible destinations.

NAP: Network Access Point. The NAP consolidates all the ISP's Point of Presence (PoP) traffic into one location. The NAP usually contains several GSR/TSR and DNS servers, as well as other name resolution services servers. ISPs do not charge fees to each other because of the assumption of symmetric traffic interchange.

NAT: Network Address Translation. NAT provides for masquerading internal private IP addresses for public IP addresses assigned by the national NIC or IANA.

NEBS: Network Equipment Building Standard. Standard in terms of environmental and mechanical features for Central Office components, building blocks, racks, and cooling and power supplies.

NHRP: Next Hop Resolution Protocol. This is a simple routing protocol.

NIC: Network Interface Card.

NIC: Network Information Center.

NMC: Network Management Center.

NMS: Network Management System. Every network needs operation, maintenance, and administration. For TelCos this is known as OA&M; in LANs this set of functions typically is carried out using HP OpenView.

NNTP: Network News Transport Protocol. NTTP is an Internet application protocol used primarily for reading and posting Usenet articles.

NPF: Network Processor Forum. Industry standardization forum for network processors, switch fabrics, and traffic managers.

OA&M: Operations, Administration, and Maintenance. Adds, changes, and moves users within the network; enacts load and performance monitoring and predictive failure analysis, as well as billing.

OAM&P: Operation, Administration, Maintenance, and Provisioning.

OC3:   Optical Carrier with 3 multiplexed STS frames (see STS frame). This term often describes the bandwidth (155.520 Mbps, 3 times the basic STS signal bit rate) and the protocol (PPP) of a given port, rather than the medium itself—so perhaps STS-3 is more accurate.

Octet: ITU-T term for byte; used within datagrams.

OEM: Original Equipment Manufacturer.

OIF: Optical Internetworking Forum. Industry standards forum for standardization of optical interconnects and backplanes/midplanes. See HSBI.

OSI: Open Systems Interconnect. This is an approach to enable vendor independent interconnectivity by establishing a communications model and the messages exchanged between peers, nodes, and clients.

OSPF: Open Shortest Path First. OSPF is a simple routing and route-optimizing scheme.

P-NNI: Private Network-to-Network-Interface.

Packet: A datagram of a variable length, with the variable- or fixed-length header in a fixed position within the datagram.

Packet Switching: Forwarding of packets based on their header information.

PAP: Password Authentication Protocol.

Pbit/s: Petabit (10^{12} or 1 trillion bit) per second.

PBX: Private Branch Exchange. This is sometimes also referred to as PABX for Private Automatic Branch Exchange.

PCB: Printed Circuit Board.

PCI: Peripheral Component Interconnect. PCI was initially an industry standard that defined the interconnectivity between the hosts' NorthBridge and add-on cards. PCI was later taken over by router manufacturers as a de-facto standard for communication devices such as low-end routers. It is succeeded by ATCA (Advanced Telecommunications Computer Architecture). It has evolved into a bus system for PCs, workstations, servers, and some Routers and Remote Access Systems. PCI is defined as a burst-mode-capable multiplexed bus with a 32- or 64-bit bus width and a cycle time of 15 ns or 30 ns. PCI therefore allows data transfer rates of 132 MByte/s (32 bit, 30 ns), 264 MByte/s (32 bit, 15 ns or 64 bit, 30 ns) or 528 MByte/s (64 bit, 15 ns) in the best case scenario. The arbitration is a little bit on the weak side. Additionally, the restriction in the number of slots reduces the deployment of PCI. This is why there is no huge confidence in PCI in the TelCo industry anymore. But there are viable solutions to both issues; PCI is best combined with I_2O in order to enable peripherals not only to transfer data into host memory, but intelligently into other peripherals.

PCI Express: PCI successor with High Speed Serial Links between a switch fabric and the add-on cards. Higher throughput, non-blocking operation, and hot-pluggable add-on cards are possible with the new standard for PCI in PCs, servers, and some telecommunication devices. ATCA is based largely on PCI Express.

PCI-X: PCI Extension. PCI-X is a 64-bit wide version of the PCI bus with reduced cycle time of 7.5 ns. PCI-X therefore allows burst mode data transfer rates of up to 1056 MByte/s in the best case.

PCS: Personal Communication Service, US version of GSM.

PHY: Physical Interface. This is Layer 1 in the ISO OSI model.

PIC: Physical Interface Card.

PIM: Protocol-Independent Multicast.

PIM-SM: Protocol-Independent Multicast—Sparse Mode.

PoP: Point of Presence. Usually the location of an ISP which houses one or more RASs, a WAN router for the uplink to the NAP, a proxy server, a content server, and the RADIUS server for access right administration and billing. The PoP is the first concentration stage.

POP3:   Post Office Protocol Version 3. This is the standard for transmitting, receiving and retrieving e-mails.

Port: Source or destination port address specified in the Layer 4 (TCP or UDP) header. The higher protocol layer uses this designator to identify the specific application or client that requires the given packet.

PoS: Packet over SONET.

PoS-PHY: Packet Over SONET Physical interface. This is one of two selectable Layer 2 interface options for the OC-3 port. The packet-oriented 8-bit parallel interface is designed to interface to the Optical PHY device.

POTS: Plain Old Telephony Service. It can be characterized as the group of services that use the analog interface between the phone and the Class 5 switch. It supports 3.1 kHz voice bandwidth, Fax Group 3 class 1 and 2 and data transfer using V.34 and V.34 plus/bis as well as x2 and 56Flex aside from the newer V.90/V.92 modems. These services use analog data streams between the phone or other analog terminal device and the Class 5 CO switch. Its signaling is inband, and it is limited to pulse dial or tone dial with MFC or MFC/R2, On-Hook and Off-Hook, as well as Hookflash. Most often, the phone companies guarantee user bandwidth up to 3.1 kHz, although mostly the usable bandwidth is up to 28800 bit/s at an analog CO switch. Using V.90/V.92 at a digital Class 5 CO switch, up to 56 kbit/s of bandwidth are available. xDSL modems are connected to user ports in the Class 5 switch so that xDSL can use the same cabling infrastructure as POTS, but xDSL traffic is filtered out. POTS is the counterpart to ISDN.

PPP: Point-to-Point-Protocol. PPP is widely used for dial-in into an ISP to connect to the Internet. It uses the E.164 address according to CCITT/ITU-T for the connection setup to the ISPs Point of Presence (PoP) or Network Access Point (NAP). While most home users will use PPP with a PC and a modem, most SoHo applications might prefer a router.

PRI: Primary Rate Interface. The ISDN PRI provides 30 channels for user data at 64 kbit/s and a signaling channel with a signaling rate of 64 kbit/s. ISDN PRI allows for channel bundling and therefore can be used for voice and data transmission. ISDN PRI routers are typically SoHo routers. See B-ISDN Q.931 and Q.2931.

Proxy ARP: The technique in which one machine, usually a router, answers ARP requests intended for another by supplying its own physical address. By pretending to be another machine, the router accepts responsibility for forwarding the packets.

PSTN: Public Switched Telephony Network. Pulse Code Modulation (PCM) and Q.931 (ISDN) as well as Q.2931 (B-ISDN) are used for trunk signaling, and Circuit Switching and Time Division Multiplexing (TDM) are used to switch and multiplex channels or "time-slots". Class 5 CO switches connect end users and PBXes to the PSTN, and Class 5 switches are connected via a hierarchy of Class 4 switches. The PSTN comprises the conventional worldwide telephone network. The PSTN is often referred to as POTS on the terminal device or customer side. See POTS for more information.

Q.2931:   See B-ISDN.

Q.931:   See ISDN.

QAM: Quaternary or Quadrature Amplitude Modulation.

QoS: Quality of Service (in ATM and IPv6).

QPSK: Quad Phase Shift Keying.

Queuing: A process in which data is temporarily stored in a directly accessible storage medium in accordance with its priority and latency requirements.

RADIUS: Remote Authentication Dial-In User Service. Together with the standards PAP and CHAP, RADIUS provides for a relatively high security of an access point for the attached users. RADIUS is defined in RCF 2058, describing in detail the Remote Authentication Dial-In User Service in its approach, the functions, and the protocols.

RARP: Reverse ARP.

RAS: Remote Access Server/System.

RBOC: Regional Bell Operating Company.

RED: Random Early Discard This mechanism determines when to discard cells or packets.

RFC: Request For Comment. Proposed Internet standards are published, discussed, and agreed upon if there is a working sample of the idea. Posted standards are binding for all members (and contributors).

RIP: Routing Information Protocol.

RMII: Reduced pin count Media-Independent Interface. RMII offers a reduced pin count for the interface between the Gigabit Ethernet MAC and the PHY compared to the standard MII.

Router: A special purpose, dedicated computer that attaches two or more networks and forwards packets from one to the other. In particular, an IP router forwards IP datagrams among the networks to which it connects.

RSVP: Resource Reservation Protocol.

RTP: Real Time Protocol.

SA: Source Address.

SAP: Session Announcement Protocol.

SAR: Segmentation And Reassembly. See AAL5.

SBPI: Serial Backplane Interface. One of th two selectable layer-2 interface options for the OC-3 port. The stream-oriented 8-bit parallel interface is designed to interface to the serial bus encoder/driver for bandwidth aggregator applications.

SCCP: Signaling Connection and Control Part (in CCS#7).

SCP: Service Control Point. Also Signaling Control Point or Signaling Conversion Point.

SDH: Synchronous Digital Hierarchy.

SDP: System Description Protocol.

Server: A server is a device that carries out remote procedure calls and centralized services for the attached clients.

SHDSL: Symmetric High Speed Digital Subscriber Line. This is a version of DSL with higher symmetric data transfer rates than standard DSL. SHDSL offers up to 44 Mbit/s of symmetric data transfer rate.

S-HTTP: see SSL.

SLA: Service Level Agreement. Typically a contract under which a service provider for a data, networking, or telephone service offers these services to a customer for a certain price, charge, or fee at a well-defined set of parameters for availability and reliability.

SLIP: Serial Line IP. SLIP is defined in RCF 1055.

SMIv2:   Structure of Management Information, Version 2.

SMP: Symmetrical Multi Processor. Most Class 4 and 5 Central Office switches are SMP architectures with distributed I/O.

SMS: Short Message Service. This service allows for text messaging in GSM and 3G cell phone standards. It has the potential to produce moderate amounts of traffic on the Internet, and is not timing-critical.

SMTP: Simple Mail Transfer Protocol.

SNAP: Sub-Network Access Protocol.

SNMP: Simple Network Management Protocol. SNMP is based on a set of rules and a set of tables defining which data to collect and how to interpret them. There is also a common set of functions defining how to transfer the results and which functions need to be taken over by the devices supporting SNMP. Alarming is clearly one of the more importing functions, since unauthorized access to LANs, networks, and devices must be avoided. SNMP is used to configure, administrate, operate, and maintain routers, bridges and gateways. Firewalls also need to be configured and monitored. Events include detected unauthorized access and exceeded error thresholds, while the general traffic statistics data collection includes, for example, the number of packets sent/received/lost, unrouteable packets, and so on. Currently, SNMPv2 (version 2) is in use.

SNMPv2:   Simple Network Management Protocol version 2.

SNMPv3:   Simple Network Management Protocol version 3.

SoHo: Small Office/Home Office. SoHo applications describe applications, which are typical for small or home offices. These include telephony and LAN interconnectivity as well as Internet access. Typically, a SoHo environment includes a few employees using a few phones and networked PCs. SoHo users very often use personal ISDN routers for their communications requirements.

SONET: Synchronous Optical Network. SONET is a North American telephone company fiber optic transmission system particularly suited for transporting STS signals.

SPEC: Supercomputer Performance Evaluation Committee/Counsel.

SSL: Secure Socket Layer.

STM: Synchronous Transfer Mode.

STP: Signaling/Service Transfer Point.

STS frame: Synchronous Transport Signal frame. STS and its frame format were developed by the telephony industry to avoid incompatible signal formats and accreting overhead with increased multiplexing rates. STS defines an atomic 810-byte frame that concentrates Operation, Administration and Maintenance (OA&M) control information into 36 bytes. A transmitter transmits STS frames as a constant bit stream without intervening pauses. Transmitting an STS frame every 125 microseconds produces an (8000 * 810 * 8) = 51.84 Mbps STS signal bit stream. This bit rate is the basis for the OC-x data rate.

TACACS: Terminal Access Controller Access Control System.

TCP: Transmission Control Protocol. The TCP session is used on top of IP to enable the applications to have a common networking socket. TCP is Layer 3.

TDM: Time Division Multiplex. Multiple digital signals are transmitted over a single line in a byte serial order, within individual timeslots. Every time slot is assigned to a sender or transmitter.

TelCo: TeleCommunications (Service Provider).

TLD: Top Level Domain.

TNM: Telecommunications Network Management.

TOS: Type of Service. Each IP datagram header includes a field that allows the sender to specify the desired TOS. In practice, few routers use TOS when choosing a route.

TSGR: Transport System Generic Requirements.

TSR: Terabit Switched Router.

UBR: Unspecified Bit Rate. ATM traffic type.

UDP: User Datagram Protocol.

UMTS: Universal Mobile Telecommunications System. UMTS is a standard for cell phone traffic and unifying services for mobile protocols. It is expected to be one of the big new contributors to bandwidth demand.

UNI: UNI is an abbreviation for the User-to-Network-Interface. It is not intended to be deployed in Inter-Carrier traffic.

UPS: Uninterruptible Power Supply.

upstream: Upstream is the direction of data flow from the client towards the server, or from the terminal to the node.

URL: Uniform Resource Locator. The URL describes in a symbolic form the address of a resource. The general layout of a symbolic address of Internet servers consists of the type of service to be used, the server name eventually preceded by the organization, and the extension. Examples are http://www.dekker.com or ftp.ibm.com among others.

VBR-NRT: Variable Bit Rate with Non-Real-Time transmission (non-CDV-bound). ATM traffic type.

VBR-RT: Variable Bit Rate with Real-Time transmission (CDV-bound). ATM traffic type.

VLSM: Variable Length Subnet Mask. This address mask separates the Network ID and the host ID in 1-bit resolution.

VoIP: Voice over IP. VoIP changes voice data into packets and then transmits them using packet networks such as the Internet. This requires a lot of signaling and messaging transfer between the VoIP gateway and the POTS network.

VPN: Virtual Private Network. VPN is essentially a virtual network within a public network, protected by encryption.

VRRP: Virtual Router Redundancy Protocol.

WAIS: Wide Area Information System.

WAN: Wide Area Network. The Internet and the worldwide telephony (POTS) networks are WANs. They are interconnected to each other via gateways and interchange access points. The POTS network is basically worldwide, and the Internet heavily relies on POTS access and on backbones shared with WAN backbones. Additionally, there must be gateways in order to accomplish Voice over IP transition into the POTS network, but also to provide a backbone for both networks. Nowadays the telephony network and the Internet rely on each other. Both are growing, but Internet has growth rates far in excess of the POTS network. Predictably, there will be a point of time at which the POTS network is just a part of the Internet.

WAP: Wireless Access Protocol.

WLL: Wireless Local Loop. This describes media access using wireless technologies such as radio frequency or infrared light transmissions.

X.25:   Secured WAN communications protocol over ISDN/POTS, defined by ITU-T.

xDSL: Digital Subscriber Line, irrespective of technology (ADSL, HDSL, MDSL, SDSL, VDSL, G.Lite).

XGMII: 10-Gigabit Media-Independent Interface.

References

SNMPv2: IETF RFC2570 & 1905.

Q_3 MIB: CCITT/ITU-T Q.811/812.

CMIP/CMIS (Common Management Information Protocol/Common Management Information Service): Q.710/711/712.

[RFC-1825] Atkinson, R. "Security Architecture for the Internet Protocol," RFC 1825(->2401prop), Naval Research Laboratory (August 1995).

[RFC-1826] Atkinson, R. "IP Authentication Header", RFC 1826 (->2402prop), Naval Research Laboratory (August 1995).

[RFC-1827] Atkinson, R. "IP Encapsulating Security Protocol (ESP)," RFC 1827(->2406prop), Naval Research Laboratory (August 1995).

[RFC-1885] Conta, A. and S. Deering, "Internet Control Message Protocol (ICMPv6) for the Internet Protocol Version 6 (IPv6) Specification," RFC 1885(->2463draft), Digital Equipment Corporation, Xerox PARC (December 1995).

[RFC-1884] Hinden, R., and S. Deering, eds. "IP Version 6 Addressing Architecture," RFC 1884hist(->2373(->3513prop)), Ipsilon Networks, Xerox PARC (December 1995).

[RFC-1191] Mogul, J., and S. Deering. "Path MTU Discovery," RFC 1191draft, DECWRL, Stanford University (November 1990).

[RFC-791] Postel, J. "Internet Protocol," STD 5, RFC 791std5, USC/Information Sciences Institute, (September 1981).

[RFC-1700] Reynolds, J. and J. Postel. "Assigned Numbers," STD 2, RFC 1700hist(->3232), USC/Information Sciences Institute, (October 1994).

[RFC-1661] Simpson, W., ed. "The Point-to-Point Protocol (PPP)," STD 51, RFC 1661std51, Daydreamer, (July 1994).

Nick McKeown, Thomas E. Anderson. "A Quantitative Comparison of Scheduling Algorithms for Input-Queued Switches."

Wolff, R.W. *Stochastic Modeling and the Theory of Queues*, Prentice Hall International., New Jersey, 1989.

Li, S.-Y. "Theory of periodic contention and its application to packet switching," Proc. of INFOCOM 1988, pp. 320–325.

McKeown, N., Varaiya, P., Walrand, J. "Scheduling Cells in an Input-Queued Switch," *Electronics Letters*, 29 (25), 2174–2175, 1993.

Obara, H. "Optimum architecture for input queueing ATM switches," *Electronic Letters*, 28th March 1991, pp. 555–557.

Obara, H., Okamoto, S., and Hamazumi, Y. "Input and output queueing ATM switch architecture with spatial and temporal slot reservation control," *Electronic Letters*, 2nd Jan 1992, pp. 22–24.

Hopcroft, J.E. and Karp, R.M. "An $O(n^{5/2})$ algorithm for maximum matching in bipartite graphs," *Society for Industrial and Applied Mathematics J. Comput.* 2 (1973): 225–231.

The ATM Forum. The ATM Forum user-network interface specification, version 3.0, New Jersey, Prentice Hall International, 1993.

Ken K.-Y. Chang, William Ellersick, Shang-Tse Chuang, Stefanos Sidiropoulos, Mark Horow-
    itz, Nick McKeown, and Martin Izzard. "A 2 Gb/s Asymmetric Serial Link for High-
    Bandwidth Packet Switches," *HotChips*, 1997.

Nick McKeown, Martin Izzard, Adisak Mekkittikul, William Ellersick, and Mark Horowitz.
    "The Tiny Tera: A Packet Switch Core," *HotInterconnects* 1996.

Balaji Prabhakar, Nick McKeown, and Ritesh Ahuja. "Multicast Scheduling for Input-Queued
    Switches," IEEE JSAC, MAY 1996.

Kun-Yung Ken Chang, William Ellersick, Shang-Tse Chuang, Stefanos Sidiropoulos, and
    Mark Horowitz.

"A 2Gb/s/pin CMOS Asymmetric Serial Link," Stanford University, Stanford, CA.

Nick McKeown and Martin Izzard. High Performance Switching (Proposal to TI, 1995).

Nick McKeown, "Fast Switched Backplane for a Gigabit Switched Router," Stanford Univer-
    sity, Stanford, CA.

Guy Redmill, July 2001. "An Introduction to SS7."

JJ (Joe) Coursolle, March 2001. "A Qualitative Review of Evolving Carrier Networks,
    2001–2005."

"SS7 Tutorial," SS8 Networks, 2002.

Rick O'Connor. "System Interconnect Today and the Road Ahead," Bus & Boards Conference,
    Long Beach, California, January 21, 2002.

S. Keshav and Rosen Sharma. "Issues and Trends in Router Design," *IEEE Communications
    Magazine*; May 1998.

John P. Ryan and Muayyad al-Chalabi. "After the Bubble: Building the New Telecom Indus-
    try," *The Ryan Perspective*, 2002.

Michael Orr. "When Network Design Meets Chaos Theory: Multilayer Switch Design,"
    Communication Systems Design, February 2003, p 25–28.

Dennis Sylvester, William Jiang, Kurt Keutzer: BACPAC–Berkeley Advanced Chip Perfor-
    mance Calculator, UC Berkeley http://www.eecs.umich.edu/~dennis/bacpac/mod-
    els/power.html.

Index

V

W